Soils

SOURCES AND METHODS IN GEOGRAPHY

Series editors

M.A. Morgan PhD
Department of Geography, University of Bristol

D.J. Briggs PhD
Department of Geography, University of Sheffield

Sediments
D.J. Briggs

Historical Sources in Geography
M.A. Morgan

Urban Data Sources
J.R. Short

Hillslope Analysis
B. Finlayson
I. Stratham

SOURCES AND METHODS IN GEOGRAPHY

Soils

David Briggs PhD
Lecturer, Department of Geography,
University of Sheffield

BUTTERWORTHS

LONDON · BOSTON
Sydney · Wellington · Durban · Toronto

First published 1977
Reprinted 1981
© Butterworth & Co (Publishers) Ltd, 1977

British Library Cataloguing in Publication Data

Briggs, D J
  Soils.   (Sources and methods in geography).
  1. Soils
  I. Title   II. Series
  631.4   S591   77-30023

  ISBN 0-408-70911-1

Printed in England by The Whitefriars Press Ltd., London and Tonbridge

FOREWORD

During recent years, geography has been undergoing considerable change, the underlying theme of which has been the adoption of a more rigorous scientific approach.
This has been reflected in the greater emphasis which is placed upon quantitative
and statistical methods of data collection and handling, in the attention given to
the study of process as opposed to the description of form, and in the use of an
inductive rather than deductive philosophy of learning.

What this means in practical terms is that the student and teacher of geography
need to be acquainted with a wide range of scientific methods. The student, both
at school and in further education, is increasingly becoming involved in projects or
classwork which include some form of individual and original research. To be
equipped for this type of study he needs to be aware of the sources from which
he can obtain data and the techniques he can use to collect and analyse this information. The teacher similarly requires a pool of empirical material on which he can
draw as a source of class exercises. Both must be able to tackle geographical problems in a logical and scientific fashion; to identify the problems in precise terms,
construct suitable explanatory hypotheses, and test these hypotheses in an objective
and rational manner.

The aims of this series of books are therefore to introduce a range of sources
which provide information for project and class work, and to outline the methods
by which this material can be analysed. The main concern is with relatively
simple approaches but reference is made to more detailed aspects where appropriate.

The reader will be expected to have a basic grounding in geography, and a working knowledge of mathematical methods is useful. Nevertheless, the level of explanation is such that, with little further reading, students should readily be able to
understand the selected themes. Consequently, the series will be of particular use
to students and teachers involved in A-level courses in geography in which project
and practical work figure as major components. At the same time, students in
higher education will find the books an invaluable guide to geographical methods.

M.A. Morgan   D.J. Briggs

ACKNOWLEDGEMENTS

I extend my sincere thanks to all friends and colleagues who have helped in the production of this book. In particular, I am indebted to Dr. Stephen Trudgill for giving up many hours of his time to talk over specific points and problems; to Mr. Frank Courtney for reading and commenting on an early draft of the text, and to the numerous students of Sheffield University who, knowingly or otherwise, provided data. No less, I am grateful to Sheila Ottewell for drawing the diagrams and to the office staff for typing the manuscript. Last, but far from least, I must thank my wife, Ann, for moral support, for trying to whenever possible spot my split infinitives, and for compiling the index.

PREFACE

The soil presents a particularly interesting focus of geographical study. It is interesting geomorphologically since soil formation is both controlled by, and itself influences, the development and nature of the landscape. It is relevant to biogeographical studies because it is closely associated with the growth of plants and the evolution of the vegetation. It is of concern in relation to man's exploitation of the environment, since it exerts a major control on agricultural activities and at the same time is affected by these activities. Thus, the soil acts as a link between various aspects of geography; it provides an opportunity to integrate a number of otherwise disparate approaches; it helps to emphasise the inter-relationships between geographical phenomena.

For a long time, however, it seems that pedological studies have largely been confined to consideration of the distribution of different soil types on a world scale. While this undoubtedly offers some chance to analyse relationships between the soil and the environment, for the student a far more rewarding approach is probably to consider more detailed relationships at a local level — on a scale to which he can personally relate. In this way it is possible to develop a clearer understanding of the *processes* which control the development of the soil, as opposed to concentrating upon relatively simple, and often misleading, spatial relationships. This involves not only the 'natural' processes of weathering and redistribution which occur in the soil profile, but also the 'human' influences of agriculture, drainage and reclamation.

The general theme of this book is therefore the investigation of the soil at a local level. It is not intended to be a definitive introduction to pedology, and it is assumed that the reader will already have a working background in physical geography. Instead, attention is focused upon the need to measure soil processes and relationships, and, arising from this the requirements of a sound scientific approach and an ability to use statistical methods to analyse and interpret data. Thus, consideration is given not only to the means of collecting information through field and laboratory work, but also to the principles of designing a project and interpreting the results.

The book should therefore be invaluable to students and teachers of A-level geography courses, particularly those in which individual or group projects and fieldwork play an important role. In addition, many students in higher education, either in geography or in related courses such as environmental studies, agriculture, soil science and botany, should find this a useful introduction to the methods and principles of soil study.

D.B.

CONTENTS

TABLES

ILLUSTRATIONS

PLATES

CHAPTER 1   INTRODUCTION

**1.1 MOTIVES AND
OBJECTIVES**

If we pick up and examine a handful of soil, it appears inert, lifeless. And yet, within every few grams of soil there may be millions of organisms, forming part of an active and complex web of life. In addition, there is organic matter — the decaying remnants of plant and animal tissues — which slowly but continuously is undergoing biochemical alteration and synthesis. Even the inorganic fractions of the soil are taking part in chemical reactions, through which plant nutrients are released and made available to roots. In other words, within the soil a whole range of processes may be operating; these processes offer an intriguing and stimulating focus of study.

We may examine these processes from a variety of standpoints. Geomorphologically, the soil forms an important part of the landscape, influencing and involved in the mechanisms of landscape development. It is similarly a major component of biogeographical systems, and is closely related to the growth and character of the vegetation. In addition, the soil is clearly vital from an agricultural point of view, and one of our main concerns is with the nature and maintenance of soil fertility. What perhaps makes the whole subject even more intriguing is the fact that the soil is spatially variable, and it thus provides the opportunity for studying a wide range of problems in a great variety of situations.

Studies of the soil nevertheless require a considerable amount of thought and planning. If we are to understand more clearly the way in which the soil works — its relationships to the environment, its potential for exploitation by man, and its inherent sensitivity to interference and misuse — we need to organise and structure our studies in a logical and scientific fashion. It is not sufficient merely to speculate about the processes going on within the soil, nor to make superficial assumptions about the implications of these processes. We need instead to find ways of measuring precisely and accurately the relevant mechanisms, and of evaluating the critical relationships between the wide variety of soil properties. The pages which follow are aimed at illustrating ways in which these studies can be carried out.

1.2 DESIGNING A PROJECT

1.2.1 The hypothesis

'I don't see much sense in that,' said Rabbit.   'No,' said Pooh humbly, 'there isn't. But there was going to be when I began it.   It's just that something happened to it on the way.'

Sadly, this situation seems to sum up many research projects.   The moral is, perhaps, simple: we need to keep a close watch on what we are trying to do to ensure that things do not go wrong on the way.   This means that we need to plan our research carefully at the outset; we need, in fact, to be clear in our minds about what we are trying to show and how best we can show it.

One of the main implications of this is that it is rarely rewarding to stumble hopefully but blindly in pursuit of some vague research objective.   If our objective is unclear, we are liable to waste considerable amounts of time and effort in trying to attain it; and we may not even recognise when we have arrived there!   To put this in more scientific terms, a successful piece of research is almost always founded upon a theory — a hypothesis — of what we expect to discover, of what we assume to be the truth.

Consequently, it is unsatisfactory merely to start off by saying, 'I want to study soil moisture' or 'what I'm really interested in is worms'.   We need rather to identify more precisely the focus of our attention, the central problem, and on the basis of our existing knowledge and understanding to construct **hypotheses** which will explain the problem.   Clearly, the more carefully we think about the problem at this stage and the more we know about it, the greater is our chance of producing a valid hypothesis and, at the end of our research, coming up with a positive solution.   Nevertheless, we must also be aware that our investigations may show that our initial hypothesis is unsound, and then of course we must be prepared to modify it to take into account our new observations.   In other words, hypothesis testing frequently involves several stages of building, reconsidering and rebuilding before we obtain a satisfactory conclusion.

The essential point is that we need, before we even start our research in earnest, to define our problem and to construct an **explanatory hypothesis** or **model**. The aim then is to test this hypothesis by collecting and analysing relevant data. The hypothesis we have built will act as a framework, therefore, in which to carry out the rest of the research.

If, however, we look at our initial hypothesis in more detail, we will often find that it consists of not one, all-embracing hypothesis, but of numerous related assumptions. If we are to test our explanatory model we need to examine each of the assumptions separately; here again, therefore, we may find that hypothesis testing requires several stages of investigation.

1.2.2 Data collection: sampling

Having defined our research hypothesis, the next step is to design a means of testing it. This involves the choice of a suitable research area, the selection of methods of analysis, and consideration of the statistical techniques which will be needed to evaluate the data we collect.

It is not intended to discuss statistical methods in this text [reference can be made to introductory texts such as those by Hammond and McCullagh (1974) and McCullagh (1974)]. Nevertheless, one brief point must be made here. Statistical analysis should not be seen as something tacked on the end of our project – an afterthought to give our results spurious scientific authenticity. If we are to come anywhere near to validating our hypothesis we will almost inevitably need to use statistical methods, and given this **it is vital to collect our data in a form which allows us to use relevant statistical procedures**. This means that we must collect enough of the right sort of information. Clearly, therefore, we need to decide upon our statistical methods before we collect our information; there is little virtue in amassing vast bodies of data which are beyond our abilities to analyse and interpret. Once more, we see here the importance of defining our research project and our explanatory hypotheses succinctly at the start of our investigation.

One of the main aspects which arises from our selection of statistical procedures is the **sample design**. Since we can rarely study the whole of the problem which faces us — we cannot collect and measure **all** the soil in our chosen area — we invariably need to base our investigations upon a **sample**. Our hope is that, although in studying this sample we are examining a small part of the total reality (the **population**) in which we are interested, we will nevertheless be able to generalise from it in order to make valid statements about the wider world.

The critical factor here is the question of the **representativeness** of the sample. In order that it may form the basis for generalisations, it must accurately reflect the population from which it is drawn; it must have been collected in a way which ensures that it is not biased towards any particular condition of the population.

The collection of representative samples is a fundamental problem in the natural sciences. In general, it is most easily achieved by the selection of a **random sample**. This ensures no bias, but it requires a great deal of care and thought; randomness is not synonymous with disorganisation. The simplest way to ensure randomness is to set up a hypothetical grid over the sample area, then choose values from tables of random numbers to represent co-ordinates. Each pair of co-ordinates defines a sample point.

A second decision which must be taken when collecting is where to sample within the soil profile. This is not so much a statistical question; instead it depends upon the aim of our investigation. In some cases we will be concerned with overall conditions in the soil profile, and will need to take specimens from the profile as a whole. In such situations a **channel specimen**, consisting of a narrow, vertical cut from the base to the top of the profile is most appropriate.

In other cases, we are concerned with soil conditions at a particular depth, or in a specific horizon (layer) within the profile. In these instances a **grab specimen** — a single, randomly chosen specimen taken from the appropriate depth — is often used. However, since soils are frequently extremely variable over small distances it may be preferable to take a **bulk specimen**. This involves the collection of several equal-sized specimens, spread randomly throughout the soil at the desired sample level. These are then mixed to give a single specimen.

The final aspect of sampling is the **size** of the sample — the number of observations which we must make. Statistical methods are available for defining the optimum sample size (see, for example, Daugherty, 1974), but these require the use of a pilot study. Owing to limitations of time, this is often impossible, and the principle must simply be to take as large a sample as feasible. It is often useful, however, to assess the likely accuracy and reliability of our sample by statistical analysis of the results; the necessary methods are explained in most introductory statistics texts.

1.2.3 Data collection: sample analysis

The third stage of designing a research project is the choice of analytical procedures. Once more, our choice is made more simple if we have previously established a clear research hypothesis. Without this we should need to consider a wide range of methods, measuring a variety of soil properties, solely on the off-chance that they might provide some relevant information.

Another factor which must be borne in mind at this stage is the requirements of the statistical methods we have previously decided to use. As far as possible, we should try to select analytical techniques which will provide data in the appropriate form. On the other hand, we may of course find that our choice of analyses is limited by availability of equipment, or the type of soil properties we have selected to study. As a consequence, the two questions of choosing suitable statistical and analytical techniques are closely related; each must be discussed with regard to the other.

1.3 FIELD AND LABORATORY PROCEDURES

Just as we need to ensure that the sample we collect represents the population from which it is drawn, so we must also be certain that the analytical results we obtain accurately reflect the soil properties we are measuring. This is partly a question of selecting methods which give precise results. It is also a question of consistency and reproducibility. Any method, no matter how precise, will still give the wrong answer if it is misused. To ensure accuracy as well as precision, therefore, we need to make sure that we use the methods under identical conditions each time we analyse our sample.

Of particular importance here is the nature of sample pretreatment. If variations in the techniques of storage, handling or preparation occur, these will almost inevitably lead to equivalent variations in our analytical results.

There are many aspects to this problem, and we can only mention a few of the ways in which consistency can be improved. As a general rule, however, the aim must always be to handle and treat each soil specimen in an identical manner.

1.3.1 Storage

Soils should normally be stored in a dry state, since this reduces the likelihood of biochemical changes over time. If possible, the material should be dried in an oven, or, if this is not possible, air-dried by spreading the soil out in a warm, dry room. It should then be stored in either tin foil or an air-tight polythene bag in a cool, dark place. In the case of properties such as moisture content, pH and nitrogen content, however, it is advisable to carry out the analyses while the soil is in its fresh, field state. If the samples need to be stored they should be wrapped immediately in an air-tight polythene bag and retained in a cool, dark cupboard.

1.3.2 Measuring out the specimen

In most cases considerably more soil is collected than is needed for any particular analysis. Thus it is necessary to take from each specimen a smaller volume on which to make the measurements. We meet here the same problems that we encountered in sampling in the field; we must ensure that the small portion we take represents the specimen from which it was drawn.

One of the more simple and yet effective means of achieving this is to 'cone and quarter' the original specimen. The soil is poured on to paper and the pile is split into four equal parts with a knife or spatula. Two opposite quarters are then taken, mixed and again poured on to the paper. These are then quartered in the same way. This process is repeated until a suitably sized portion is obtained.

An alternative technique is to use a sample-splitter or riffle. These can be bought commercially, or they can readily be constructed from simple equipment. Thus a number of round holes may be cut in the bottom of a cardboard box or seed tray. At one end of the box a slit is then made, extending the width of the box but slightly above the base. A length of card is inserted in this so that it covers all the holes in the base, giving a false bottom. The soil is then emptied into the box and thoroughly stirred. The false bottom is then removed and the soil allowed to filter through the holes on to a sheet of paper (or through funnels into beakers). In this way an immediate split into a large number of small portions can be obtained. For greater accuracy several of these may be remixed and split again.

Once a specimen of suitable approximate size has been obtained, it can be weighed out for the analysis. Again, however, care must be taken. If the soil is simply poured on to the balance, there is the likelihood of sorting taking place, and bias will result. Instead, it is necessary to spoon out the soil with a small scoop or spatula. It must always be remembered that unless we ensure that the material we measure — which in the end may be no more than a few grams selected from a whole field — is representative of its parent population, our results are almost meaningless.

1.3.3 Analysis

Finally, of course, we must ensure that we use the same procedures during analysis. In this context it is often valuable to repeat measurements on at least a selection of specimens, in order to check that results are consistent. In many cases it is also a good policy to prepare replicate specimens for analysis, each of which can then be measured, and the average value taken from the results for each specimen, after omitting any 'rogue' values. Apart from ensuring that mistakes in our analysis are spotted in this way (we are hopefully unlikely to repeat the same mistake on several specimens), this has the advantage of allowing us to assess the inherent reproducibility of the results.

1.4 SECONDARY SOURCES

Although it is probably true that the most interesting and rewarding source of information about the soil comes from studying the soil itself, there are nevertheless several secondary sources which can provide valuable data. Possibly the most widely available are the soil maps which are published by the soil surveys of England and Wales and of Scotland (see Appendix 2 for addresses). As yet a relatively small proportion of the United Kingdom has been mapped by these organisations, but even so a wide range of maps can be obtained, illustrating the soil types found in a variety of situations. In addition, with each map a record or memoir is available which describes the soils and land use of the area in some detail.

These maps are available at two scales (1:63,360 and 1:25,000), though only the 1:25,000 maps are still being produced. Each provides the basis for a number of investigations. We may, for example, use these maps to study the relationships between soil and geology, relief or altitude. Similarly, we can use the soil maps to assess the influence of soils upon land use; for all the areas covered by the 1:63,360 soil maps, land utilisation survey maps are available. These were published in the 1930's and are to some extent out of date now. However, a more recent survey of land use has been carried out at a scale of 1:25,000 by the Second Land Utilisation Survey, and in a few areas the published maps coincide with the 1:25,000 scale soil maps.

Several other related maps have also been produced. Of particular importance are the land use capability surveys carried out both by the Soil Survey of England and Wales and by the Ministry of Agriculture, Fisheries and Food. The former are available for many of the areas mapped at 1:25,000 scale as part of the soil coverage; the latter are at a 1:63,360 scale and are available for almost the entire country. Both surveys provide a classification of the land on the basis of its agricultural potential. Although it would be misleading to compare these at a simple level with the soil maps — since the soils data are frequently used to provide the land use capability classification — they do allow us to consider the importance of particular soil conditions in determining the agricultural value of different areas.

Although we have discussed these secondary sources of information mainly in the context of England and Wales, to a great extent similar maps are available for Scotland and parts of Ireland. A list of relevant addresses for the whole of the United Kingdom are provided in Appendix 2.

In addition to providing the basis for fairly straightforward comparisons of soil and environmental factors, or soil and land use, the soils maps also offer scope for more open-ended discussions. Thus, they act as a foundation for consideration of a variety of planning problems such as the question of locating footpaths or nature trails in upland areas in a way which will cause the least problem of soil erosion. We may similarly use soil and land-use capability maps to consider the problems of planning urban or industrial developments, roadways and so on, in a way which will minimise the loss of fertile agricultural land.

In all these examples, the obvious advantage of using soils maps is that they allow us to carry out soil studies without ever getting mud on our boots. Nevertheless, a certain degree of caution must always be employed. In the first place, it is important to remember in the more open-ended and unstructured exercises that soil is only one of a host of important factors which must be taken into account. Additionally, we must beware of simplistic interpretations about soil patterns and soil relationships based upon maps. The soils map, after all, is not an accurate reflection of reality; rather it is the surveyor's impression of that reality. And the surveyor may well have been influenced by the very factors with which we are trying to compare the distribution of soils. For example, in trying to identify the boundary between two soil types the surveyor may well decide to place it in accordance with a break in slope, or a change in the land use. He may well feel that these changes reflect the change in the soil conditions which his fieldwork has shown exists. Small wonder, however, that when we analyse the distribution of soils in that area, we find a relationship between the soil and topography, or the soil and land use.

CHAPTER 2 PHYSICAL PROPERTIES

2.1 INTRODUCTION

The soil is commonly described as a **three phase system**, composed of solid, liquid and gaseous phases. In most soils, the solid phase makes up the vast majority of the soil mass, and over half its volume (*Figure 2.1*). It consists of **mineral matter** derived from the breakdown of rocks, and **organic matter** from the decomposition of plants and animals. The liquid phase is composed predominantly of water, enriched with dissolved solids; the gaseous phase of air, enriched with carbon dioxide from the respiration of soil animals and plant roots.

Physically, these phases can be considered as an amalgam in which the solid particles are packed, somewhat loosely together, leaving between them voids which are filled with air and water. The character of the solid particles, and the way in which they are packed together determine the physical properties of the soil. Thus the physical conditions of the soil involve two related aspects — the nature of the individual solid particles and of the aggregates which these form.

2.1.1 Soil texture

The mineral particles which make up the bulk of the solid phase of the soil can be described in various terms. They can be categorised according to their mineralogy — whether they are quartz or feldspar, montmorillonite or kaolinite; they can be classified according to their shape — their roundness or sphericity. They can also be described in terms of their size.

All three properties are important in affecting the processes and character of the soil. The mineralogy, for example, influences soil fertility through its control on the type and quantity of plant nutrients which may be released by weathering. The shape of the particles affects their packing and thus is related to soil structure. But particle size is probably the most important, for it is closely related to a wide range of other properties and processes — to structure and drainage, to root growth and plant nutrition, to weathering and soil formation. It is, in addition, relatively easy to measure, and thus is of widespread use in the study of soils.

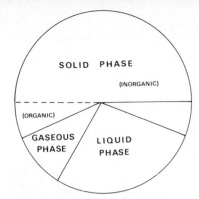

Figure 2.1 Soil as a three-phase medium: the proportions of solid, liquid and gaseous material in a typical mineral soil, expressed on the basis of volume

Soil is composed of mineral particles which differ greatly in size. At one extreme are the large rock fragments and boulders created by physical disintegration of bedrocks; at the other extreme are the minute clay particles which are produced by prolonged chemical alteration of soil minerals. Differences in the nature of the rocks from which the soil is derived, or in the processes of weathering which produce the soil, result in differences in the range of particle sizes in the soil.

In the field, the differing sizes of the constituent particles give each soil a characteristic feel (*Table 2.1*). Thus, a soil composed mainly of coarse particles feels light and gritty; one composed mainly of clay feels heavy and sticky. The feel of the soil is often referred to as its **texture**. Hence, a farmer talks about soils with a heavy texture, or about loamy soils, and so on. In reality, these descriptions refer to more than just the size of the soil particles, because properties such as the moisture content and organic content also influence the feel of the soil. Nevertheless, the size of the mineral grains is normally the dominant factor, and it is this which we measure when we assess the texture of the soil.

2.1.2 Soil structure

The individual mineral grains which make up the solid phase of the soil do not occur as loose particles, but are bound together as aggregates. These aggregates — or **peds** as they are often known — give the soil a structure, an architecture. They also have numerous implications for soil processes. They give the soil a chemical and physical stability considerably greater than that of loose sediments. They help control water movement through the soil, allowing excess water to drain away, but retaining moisture in the small pore spaces within the aggregates. Moreover, the architecture of the soil produces a range of environments which differ in terms of their aeration, humidity and temperature. These conditions provide a diversity of habitats, suitable for many different soil organisms; thus the structure has important ecological effects.

The structural properties of the soil depend ultimately upon the character of the individual grains and the way in which they are held together in peds. These peds occur for a variety of reasons. Clay particles and organic matter in the soil

TABLE 2.1 GUIDE CHART FOR HAND-TEXTURING OF THE SOIL

Textural class	Description
Sand	Gritty, non-coherent; does not form a cube even when moist; leaves fingers clean
Loamy sand	Gritty; forms a weak cube when moist; slightly coherent
Sandy loam	Gritty; markedly coherent; forms a firm cube and will roll into a weak thread; sand grains can be seen and are audible when the soil is moulded; does not polish
Sandy silt loam	Silky and somewhat gritty; coherent; forms a cube and may roll into a weak thread; leaves a silty film on the fingers; does not polish
Sandy clay loam	Gritty and somewhat sticky; resists deformation; will form a firm thread; sand grains can often be seen and heard when the soil is moulded; may polish
Clay loam	Sticky; will polish quite readily; will form thread but will not form a firm ring
Silt loam	Silky; will not polish and forms only a weak thread; leaves a silty residue on the fingers
Silty clay loam	Silky and somewhat sticky; polishes; will form a thread but will not form a firm ring; leaves a silty residue on the fingers
Silty clay	Sticky and somewhat silky; resists deformation; will form a thread and may form a weak ring; leaves a silty residue on the fingers; polishes readily
Sandy clay	Sticky and somewhat gritty; sand can often be seen but is not audible when the soil is moulded; will polish, but sand grains may scar the smooth surface; will form a thread but will not form a firm ring
Clay	Sticky; resists deformation; polishes readily; forms a thread and this can easily be bent to form a firm ring

have a natural **cohesion**, owing to the existence of minute electrochemical charges on their surfaces which cause mutual attraction. In addition a variety of agents bind the soil particles together. Chemical cements, such as calcium carbonate, and organic gums produced by soil animals and decaying plant material, stabilise the aggregates produced by the initial attraction of the particles for each other, while plant roots and the thin threads of soil organisms such as fungi and actinomycetes, enmesh the particles. Thus the soil structure is closely related not only to the texture of the soil, but also to its chemical and organic content.

2.2 TECHNIQUES

2.2.1 Soil texture

Introduction

As we have seen, the texture of the soil depends largely upon the size of the soil particles. It has also been noted that many people use somewhat vague terms such as 'heavy', 'loamy' and 'light' to describe soil texture. For accurate, scientific analysis of the soil these descriptions are inadequate; they lack precision, they may be ambiguous, and they are not based on any measurable, reproducible criteria which ensure that the descriptions are consistent. To determine soil texture, we therefore need a more strict and logical means of measuring particle size.

Material between about 0.05 and 5.0 mm can be analysed by **sieving**; a sample is passed through a series of sieves of different mesh size and the proportion retained on each sieve is calculated. For even finer materials (<0.05 mm), however, this method is unsuitable and measurement is commonly made by **sedimentation** techniques. These are based upon the principle that the rate at which a particle settles from a suspension is proportional to its diameter; large particles settle more rapidly than small ones. Thus, by measuring the time it takes particles to settle from a suspension of water it is possible to estimate their diameter.

In reality, neither sieving nor sedimentation techniques measure the diameter of individual grains, but merely subdivide the soil into several size fractions. By detailed analysis it is possible to subdivide the soil sample into a large number of fractions in this way, and thereby obtain a very fine and precise estimation of particle size. This level of detail is essential if subtle differences in soil texture are being investigated, and it enables a number of particle size parameters to be calculated which act as numerical descriptions of soil texture. For example, the results from particle size analyses can be plotted as **histograms** or **line graphs** to show the distribution of particle sizes in the sample. In many cases they are drawn up as a **cumulative percentage curve**, using probability paper, and from this estimates of parameters such as **mean size** or **sorting** can be made. This approach is frequently used in the analysis of sediments (see the companion book in this series: *Sediments*, Briggs, 1977). However, this method has several disadvantages. In particular, it requires detailed and lengthy analysis which makes it unsuitable for routine analysis of large numbers of samples, or for 'on-the-spot' analyses by farmers, agricultural advisors and so on. Consequently, an alternative, simpler method of textural analysis is often employed.

Only the material finer than 2 mm in diameter — the **fine earth** fraction as it is called — is normally analysed. This is considered to be the part of the soil which is both physically and chemically active; the coarser material, formed of mineral aggregates and rock fragments rather than individual grains, contributes less to soil processes and plant growth. The fine earth fraction is subdivided into three grades: **sand**, **silt** and **clay**. There is no universally agreed definition of these three grades, but in Britain they are normally defined thus:

Sand   2 mm–0.06 mm (2000–60 μm)
Silt   0.06 mm–0.002 mm (60–2 μm)
Clay   <0.002 mm (<2 μm)

Note that the particle size can be expressed in terms of **micrometres** symbolised by μm. One micrometre is one-thousandth of a millimetre (1 μm = 0.001 mm), and since we are often dealing with particles only a fraction of a millimetre in diameter, the use of micrometres is valuable because it avoids the need to use fractions or decimal points.

Textural analysis of the soil therefore consists of estimating the proportion of the sample in each of the three size grades; the texture can thus be expressed in terms of the percentage of sand, silt and clay. In many cases, however, on the basis of these values, the soil is assigned to a **textural class**. To do this use is made of a **triangular texture diagram** (*Figure 2.2*). This consists of a triangle, each side of which relates to a single size grade and is graduated from 0 to 100%. The triangle

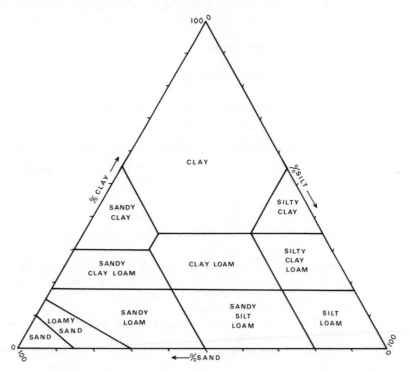

Figure 2.2 The triangular texture diagram. The diagram distinguishes textural classes on the basis of their contents of sand, silt and clay

is subdivided into a number of textural classes, such as clay, clay loam, silty clay loam and so on. Using the percentages of sand, silt and clay any sample can be located on this diagram and its textural class identified.

In some instances the textural diagram may be used in a rather different fashion. Using simple techniques such as **hand texturing**, the soil may be located on the diagram, and then the approximate percentages of sand, silt and clay read off. This is possible because each texture class — each combination of sand, silt and clay — has a characteristic feel. This is the basis of hand texturing of the soil.

Measurement

HAND TEXTURING

Equipment: Soil texture diagram

Sample: Small, moist clod

Procedure: Moisten sample until moisture glistens on surface.

Rub between fingers and assess the grittiness, smoothness (silkiness) and stickiness.

Try to form a cube, thread and ring from the sample; try to polish the surface.

Assign to the appropriate texture class.

The principle of hand texturing is that the three main size grades (*Table 2.1*) have a different feel when rubbed between the fingers. **Sand feels gritty** and, when present in large amounts, can be heard to rasp as the particles are rubbed together. **Silt feels silky** and smooth; when dominant in the soil it gives a distinct fluffy or feathery feel. **Clay is sticky**, tenacious and plastic; the particles adhere to each other as well as to the skin, so that when the fingers are drawn apart the clay resists deformation.

The problem, however, is to determine the relative proportions when all three size fractions are present. In such cases the characteristics of the individual fractions are often masked. Consequently, several different tests need to be carried out, using a moistened soil sample. One of particular use is to rub the thumb over the moist soil to determine whether it leaves a smooth, polished surface. This enables the distinction of soils with moderate amounts of clay (20% or more). Another useful test is to try to form a cube from the soil. Soils with about 5% or more clay form reasonably cohesive cubes. A further test is to roll the soil into a thread. Again this is only possible if clay is present in significant quantities (10–15% or more). Where clay is abundant (over 25%) the thread can be bent into a ring; the higher the clay content the firmer will the ring be.

A simplified scheme of tests is given in *Table 2.1*. It should nevertheless be remembered that factors other than the sand, silt and clay contents may influence the feel of the soil. Organic matter often makes the soil feel rather smooth and silt-like; dry soil often feels coarser than it really is owing to incomplete separation of the particles. Consequently, there is no substitute for practice. Ideally, samples of different soils should be collected, analysed if possible, and hand-textured; every so often the consistency of our classifications can be checked by referring back to these type samples.

HYDROMETER ANALYSIS

Equipment: Mortar and pestle
2 mm sieve and pan
1000 ml measuring cylinder
500 ml beaker
Bunsen burner, tripod and gauze
Dessicator containing silica gel or calcium chloride
Thermometer
Oven
Balance
Hydrometer (direct reading, from Gallenkamp)

Reagents: Hydrogen peroxide (20–30%)
Calgon (sodium hexametaphosphate)

Sample: *c*. 100 g soil

Procedure: a) **Pretreatment** (this can be omitted if the organic content is less than 5%).

Pour *c*. 100 g soil into 500 ml beaker; slowly add 100–200 ml 30% hydrogen peroxide and stir. (N.B. this solution is highly caustic.)

Heat gently over a bunsen burner and increase heat to boiling point.

Simmer until most of the hydrogen peroxide has evaporated but do not allow the sample to dry, otherwise an explosion may occur.

Place the sample in an oven and dry at 105°C for 24 h.

Cool in a dessicator.

b) **Measurement**

Sieve the dry soil through the 2 mm sieve and collect the fine earth fraction in the pan.

Pour into the pestle and gently grind using a rubber tipped mortar.

Weigh out up to 60 g ground soil and record the weight as **W**.

Pour the soil into the 1000 ml measuring cylinder; top up to 1000 ml with water and add 1.0 g Calgon.

Stir thoroughly and leave for several hours to reach room temperature.

Measure the temperature with the thermometer; invert the cylinder several times, covering the end with a hand to avoid spillage, until the contents are thoroughly mixed; leave in a cool, shaded place.

After 20 s, gently insert the hydrometer; allow to settle; then 50 s after stirring, read the scale on the stem at the top of the meniscus; remove the hydrometer carefully and record the reading as R_1.

Repeat the procedure after 8 h; record the value as R_2.

Calculation: Adjust the hydrometer readings by adding 0.3 g for every 1 deg C above 20°C, or subtracting 0.3 g for every 1 deg C below 20°C. Enter the corrected values into the following equations to give percentages of sand, silt and clay.

$$\% \text{ Sand } = \frac{W - R_1}{W} \cdot 100$$

$$\% \text{ Silt } = \frac{R_1 - R_2}{W} \cdot 100$$

$$\% \text{ Clay } = \frac{R_2}{W} \cdot 100$$

(If an original sample of 60 g is used, $W = 60$)

Various methods of size analysis based upon the principle of sedimentation have been devised. This principle is stated in **Stokes' Law: the velocity of a particle falling through a viscous medium is directly proportional to the diameter of the particle**. Thus, by measuring the velocity of fall of the particles, it is possible to estimate their diameter.

In reality it is difficult to monitor the fall of individual particles through a suspension. Instead, **the hydrometer technique measures the change in the density of the suspension over time**. As particles settle from a soil–water suspension, the density of the remaining suspension decreases. Since large particles fall most rapidly, they are the first to settle out. Sand sized material, for example, will have settled within about 50 s. Finer particles take longer; clay may require well over 8 h to settle from suspension. Thus, by measuring the change in the density of the suspension it is possible to assess both the diameter of the particles left in suspension and the proportion which they form of the total sample.

Although the hydrometer method is an extremely accurate yet simple technique of textural analysis, there are several sources of error and difficulty which need to be considered. In many ways the prime source of error lies not in the actual technique, but in the pretreatment of the sample. In order to measure the size of the individual soil particles it is necessary to remove the organic matter and disaggregate the soil. Removal of organic matter is only necessary in soils containing a noticeable (i.e. 5% or more) quantity of humus or plant debris. Removal is achieved by treating the sample with dilute hydrogen peroxide. Disaggregation is generally carried out by grinding the sample in a mortar with a rubber tipped pestle, or by dispersing the soil in a solution of Calgon and water (1 g/l). The former method is the simplest but the most drastic, and it may lead to destruction of individual particles. Dispersal in Calgon is less likely to damage the particles but may be inadequate to ensure complete disaggregation. The problem, therefore, is to attain complete disaggregation without breaking up any of the mineral grains.

This problem, of course, is not confined to the hydrometer method of analysis; it is common to all techniques of particle size analysis. Nevertheless, because of the difficulty in pretreating the sample, it is probably impossible to achieve an accuracy of better than ±2% in the measurement of clay contents of the soil.

The other main problems are associated with the insertion and reading of the hydrometer. It is essential to exert extreme care when lowering the hydrometer into the suspension and when raising it. On both occasions some disturbance of the suspension occurs, and it is necessary to minimise this as much as possible. On some occasions, too, a froth may develop on top of the suspension, normally caused by the addition of too much Calgon. This can be dispersed with a few drops of amyl alcohol or acetone.

2.2.2 Aggregate stability

Introduction

As we have noted, the individual particles of the soil are generally arranged in aggregates, or peds. One of the major characteristics of these peds is their shape (*Figure 2.3*); another is their stability. The aggregates tend to be less stable when wetted than they are when dry, a phenomenon which arises for two main reasons. Water

GRANULAR
Mull horizons, cultivated loamy soils

BLOCKY (SUB-ANGULAR ROUNDED)
B horizons in loamy soils, ploughed
horizons in loamy-clayey soils

COLUMNAR
B horizons in clayey soils

PRISMATIC
B horizons in clayey soils
especially those containing
montmorillonite

PLATY
B horizons in compacted clayey
soils or laminated B/C horizons
(e.g lacustrine clays)

Figure 2.3 Types of soil structure

entering the aggregates interferes with the electrochemical forces which cause the particles to cohere to one another; it also dissolves some of the cements which bind and stabilise the aggregates.   The effect, however, is not identical in all soils.   It varies according to properties such as the amount and type of clay, the organic matter content, and the type of cement in the soil.   Thus, some soils are able to withstand the effects of wetting more effectively than others.

The resistance of the soil to breakdown upon wetting — its **aggregate stability** as it is known — is important for a variety of reasons.   Soils with low aggregate stability break down rapidly under the influence of rainfall, and, as a consequence, they tend to form compact surface crusts which prevent infiltration of the water.   The rainfall is therefore unable to drain into the soil, and instead it flows over the surface, as overland flow.   On sloping land this may lead to serious soil erosion, as the water picks up and carries away the individual particles which have developed as a result of the destruction of the soil aggregates (*Plate 1*).   Formation of crusts in this way also hinders plant growth, for seedlings are unable to penetrate the compact surface layer, and thus a large proportion of plants fail to reach maturity.

Measurement of aggregate stability therefore gives a direct measure of the susceptibility of the soil to structural deterioration under the influence of rainfall.   Indirectly it also has wider significance, for soils with a low structural stability tend to be prone to physical damage by farm machinery and trampling by animals (*Plate 2*). The use of tractors or heavy cultivators or harvesting machines on such soils may destroy the aggregates and cause surface compaction. This occurs particularly when the soil is wet, for it is then that the stability of the aggregates is at a minimum. Damage of this type has a number of consequences. It hinders root growth, it slows down water movement and it reduces the rate of air circulation. All three effects tend to reduce plant growth.

Measurement

Equipment: Garden sieve
1000 ml graduated beaker
Sample: Large clod

Plate 1. Gullying caused by erosion of soil on compacted reclaimed land. Because the porosity of the soil is low, water is unable to percolate through the soil. In heavy rain it rapidly becomes saturated and water runs over the surface, collecting in small streams and eroding gulleys (Photo J. Owen).

Plate 2. Poaching — a result of trampling of wet soils by animals. When cattle are grazed during winter on soils of low aggregate stability considerable structural damage may result (Photo D. Briggs).

Procedure: Place the soil in the sieve and gently agitate.

Pour the soil retained on the sieve into the beaker and gently tap it so that the soil settles; note the volume of the soil and record as V_1.

Carefully fill up the beaker with water, avoid damaging the aggregates by playing water directly on to the peds.

Allow to stand for 30 min, then carefully pour off the water.

Tap the beaker gently; note the volume of the soil and record as W_2.

Calculation: Calculate the aggregate stability of the soil as

$$S(\%) = \left(1 - \frac{V_1 - V_2}{V_1}\right) . 100$$

A number of methods have been devised for measuring the aggregate stability of the soil. Possibly the most widely used approach involves shaking up a sample in water, then wet sieving it. The proportion of the initial sample remaining as aggregates coarser than 2 mm in diameter provides a measure of the aggregate stability.

The same principle can be more simply applied. The volume of the dry aggregates can be determined, then the soil can be saturated and the volume again assessed. The change in volume, as the aggregates break down and settle more closely together, gives a measure of the aggregate stability. A completely stable soil will not change in volume during this procedure and will have a stability of 100%. Although this approach is less precise than the sieving method it has the advantage of speed, so that a number of replicate samples can be analysed and their average reading taken.

2.2.3 Bulk density

The bulk density of the soil is the ratio of its mass to its volume. Thus it is expressed by the relationship:

$$\text{Bulk density } (D_b) = \frac{\text{Weight } (W)}{\text{Volume } (V)}$$

The two main factors affecting this relationship are the composition and packing of the soil. The composition of the soil solids — in particular the relative proportions of mineral and organic matter — affect the intrinsic density of the soil materials. The packing controls the size and number of voids within the soil.

Most of the minerals which make up the soil — quartz, feldspar, mica and clay minerals — have a density of about 2.6 to 2.7 $g\,cm^{-3}$. Organic matter, on the other hand, has densities of less than 0.4 $g\,cm^{-3}$. Consequently, as the organic content of the soil increases, its bulk density tends to fall. Whereas the average bulk density for mineral soils is therefore about 1.25 $g\,cm^{-3}$, many peats (composed almost entirely of organic matter) have bulk densities of 0.5 $g\,cm^{-3}$ or less.

In many cases, however, it is variations in the packing of the soil particles which are more important. Thus, mineral soils may show a range from 1.0 to 2.0 $g\,cm^{-3}$, as a result solely of differences in the volume of voids (the void ratio or porosity).

Low values are generally associated with recently cultivated, well-structured topsoils, in which abundant pore spaces exist both within and between the aggregates. High values occur where these pore spaces have been destroyed by compaction and consolidation. Thus, high bulk densities are characteristic of subsoils which have been compacted by agricultural activity.

Measurement

CORE METHOD

Equipment:
: Core tube or sampling box
  Hammer and block of wood
  Balance
  Oven
  Dessicator

Sample:
: Core sample

Procedure:
: Collect sample by hammering the core tube or sampling box into the soil, using the block of wood as a buffer.

  Calculate the volume of the sample (V) by measuring the depth of the hole (h) and radius (r); volume = $h\pi r^2$.

  Remove soil from the sample tube and dry in an oven at 105°C for 24 h.

  Cool in a dessicator; weigh and record weight as W.

Calculation:
: The bulk density (D_b) of the soil is given by the dry weight divided by the volume of the sample:

$$D_b = \frac{W}{V}$$

SAND BATH METHOD

Equipment: 500 ml beaker
500 ml graded, clean sand
Balance
Oven
Dessicator

Sample: Small clod of soil

Procedure: Fill the beaker to the 500 ml marker with dry sand, tapping gently to aid settling. Weigh the sand and calculate the weight of 1 cm^3 of sand.

Dry the soil sample in an oven at 105°C for 24 h; cool in the dessicator; weigh and record weight as **W**.

Sprinkle a thin layer of sand in the bottom of the beaker; place the soil clod in the beaker and pour sand around it, tapping the beaker to aid settling.

Fill the beaker to the 500 ml level with sand. Weigh the remaining sand (left from the original 500 ml) and calculate its volume; this gives the volume of the soil.

Calculation: The bulk density of the soil sample is given by the weight of the soil divided by the volume of the displaced sand:

$$D_b \ (\text{g cm}^{-3}) \ = \ \frac{W}{V}$$

Possibly the simplest method of measuring bulk density is the core technique. A core tube (or sampling box) is hammered into the soil, using a block of wood as a buffer to avoid damaging the sampler. The volume of the sample can be calculated either by measuring the height and radius of the soil in the core tube or, preferably, by measuring the dimensions of the hole from which it was drawn; the latter is preferable since it is often found that insertion of the core tube causes compaction of the soil sample. The soil can then be dried and weighed, and its bulk density calculated.

Core tubes for this method can easily be produced from metal tubing with an internal diameter of about 5 cm or more; an overall length of 10–15 cm is suitable, and the cutting end should be ground to give an angle of approximately 45°. If possible a marker should be scribed on the side of the core denoting a suitable volume (e.g. 100 cm^3). The core tube can then be inserted to this marker on each occasion and the volume of the sample is automatically known.

An alternative approach is possible using individual aggregates or clods of soil. Their volume is calculated by the displacement method. The volume of sand displaced from a beaker by the soil sample is calculated, and from this the bulk density of the soil can be assessed. The advantage of this method is that compaction or damage to the soil during sampling can be minimised. The method is only suitable for fairly cohesive, clayey soils, however. In addition, where a single aggregate is used, measurement refers only to the bulk density of that aggregate; it does not include the voids between the aggregates. As a result, this method tends to give rather higher bulk densities than the core method. Because of this difference it is important when carrying out measurements of bulk density to describe the technique used.

2.2.4 Porosity

Introduction

Soil porosity is closely related to bulk density. It refers to the volume of pore spaces within the soil, and thus is inversely related to the density of particle packing. Two types of pore spaces occur: small voids exist between the soil particles as a result of the imperfect 'fit' of the particles together, while larger voids occur between the aggregates.

These pore spaces are important for several reasons. They act as the main passages for air and water movement through the soil, and thus control aeration and drainage. They also provide space in which soil organisms can live, and into which plant roots

TABLE 2.2 CLASSIFICATION OF SOIL POROSITY

Abundance		Size	
Description	Number/100 cm²	Description	Diameter (mm)
Rare	< 5	Very fine	<0.5
Few	5–10	Fine	0.5–1
Common	10–100	Medium	1–3
Abundant	>100	Large	3–5
		Very large	>5

can extend. Through these relationships they influence the chemical conditions of the soil.

To a great extent it is not the total volume of the pore spaces which is important, but their diameter. The radii of the pores govern the force that retains water in the soil; this force is greatest in small pore spaces and least in large pores. Similarly, the size of the pores determines the ease with which plant roots can extend into the soil; free entry of roots is only possible in the largest pores (0.3 mm or more in diameter). Unfortunately, measurement of pore size cannot easily be carried out. It is possible, using a hand lens or a low-powered microscope to study a small clod of soil, to classify the pores on a rather simple scale (*Table 2.2*), but more accurate measurements require complex methods such as studies of thin sections of soil.

Although of less significance, the total volume of pore spaces — the porosity — is easier to analyse. This is normally assessed indirectly, by measuring the bulk density (D_b) of the soil and the average **specific gravity** (G_s) of the soil particles. The porosity, expressed as the percentage of the soil volume occupied by voids, is then given by the relationship:

$$P\ (\%) \ = \ \left(1 - \frac{D_b}{G_s}\right). \ 100$$

Because of the close relationship between porosity and bulk density, the highest porosities are normally found in soils of loamy texture — in clay loams, sandy silt loams and sandy loams. In these the combination of sufficient large particles to provide abundant coarse pores within the aggregates, and sufficient cohesion to produce a fine, yet stable structure, results in a high total porosity. In clayey soils the volume of pores within the aggregates is reduced because of the fine, closely packed nature of the clay particles; in sandy soils, the lack of cohesion leads to a reduced volume of voids between the aggregates.

Measurement

Equipment: Equipment for bulk density measurement as above

Pycnometer jar or specific gravity bottle

Mortar and pestle

Sample: *c.* 100 cm^3 undisturbed soil

Procedure: Calculate the bulk density of the soil, as above; record the value as D_b.

Dry *c.* 100 g soil in an oven at 105 °C for 24 h.

Cool in a dessicator; weigh and record weight as W_1.

Gently grind the soil in a mortar.

Fill the pycnometer or specific gravity bottle to the outlet in the top with water; wipe away any excess water from the outside of the jar; weigh and record the weight as W_2.

Take off the lid, pour half the water away and pour the soil into the jar; stir thoroughly to remove all the air bubbles.

Replace the lid; top up the jar with water; cover the hole at the top of the jar with the hand and shake thoroughly to remove any further air bubbles; top up again if necessary and repeat until all the bubbles have been removed; wipe the outside of the jar.

Weigh the jar and its contents; record weight as W_3.

Calculation: The specific gravity of the soil particles is given by the equation

$$G_s \; (\text{g cm}^{-3}) \; = \; \frac{W_1}{W_2 + W_1 - W_3}$$

The porosity is given as

$$P \; (\%) \; = \; \left(1 - \frac{D_b}{G_s} \right) 100$$

(N.B. In very sandy, siliceous soils containing little organic matter, it is justifiable to assume a specific gravity of 2.65 g cm^{-3}; in this case porosity can be estimated directly from measurement of bulk density.)

The technique of porosity measurement depends upon the character of the soil. In very sandy, siliceous soils, containing little or no organic matter, the specific gravity of the soil particles approaches 2.65 g cm^{-3} (the specific gravity of quartz). Thus, only the bulk density needs to be calculated. Frequently, however, the soil contains sufficient clay and organic matter to make separate analysis of the specific gravity necessary. This is most easily achieved by use of a pycnometer jar or specific gravity bottle. The jar is filled with water and weighed; some of the water is then replaced by soil of a known weight and, all the air bubbles having been removed, the jar and its contents are again weighed. In this way the volume of the soil particles can be estimated and their specific gravity (weight/volume) assessed. Knowing both specific gravity of the soil particles, and the bulk density of the undisturbed soil, we can then determine the porosity.

The main errors in this analysis are either in the determination of the bulk density, as explained above, or in the weighing of the pycnometer and soil. In particular, inadequate stirring after the soil has been added leaves air bubbles in the pycnometer and these reduce the weight. The effect is to give a higher value of specific gravity than the true value, and thereby lead to an overestimation of porosity.

2.3 PHYSICAL PROCESSES

2.3.1 Physical weathering

Freeze–thaw

One of the main processes of physical weathering is freeze–thaw activity. As the temperature falls below 0 °C, water held in the soil and surface layers of rocks freezes; as it does so its volume expands by about 10% and considerable stress is set up, forcing the soil particles apart. When the temperature rises, the water melts and its volume diminishes. New water can now enter the pore-spaces and they are thus refilled. Upon refreezing expansion again takes place and the soil particles are forced further apart. This results in the slow breakdown of individual mineral grains and in the disintegration of soil aggregates.

Although in Britain, freeze–thaw activity occurs only on a relatively few days during winter, its effect on soil structure may be appreciable. Indeed, it is for this reason that farmers tend to plough their fields during the autumn and leave them fallow over the winter; the winter frosts help to break down the aggregates and produce a fine, even tilth. In addition, many soils in Britain show a legacy of freeze–thaw processes which operated during the Ice Age. Repeated activity has led to complete disaggregation of the soil and produced compact, structureless layers, known as duripans or fragipans.

Wetting-and-drying

Even without freezing, water can set up considerable stresses within the soil and profoundly affect soil structure. In clay soils, particularly, the absorption of water by **colloidal** particles causes expansion and reduces the strength of the forces which bind the particles together. Consequently, the aggregates are weakened. In addition, as the aggregates expand they tend to move in relation to each other, one block slipping over the next. This leads to the development of polished surfaces to the aggregates — slickensides as they are frequently known.

By contrast, during drying the colloids lose water and contract. The aggregates are strengthened, for the particles are pulled closer together. At the same time the contraction of the aggregates causes large fissures to develop along the lines of weakness. Under certain conditions these take the form of deep polygonal cracks. The effect of this is to produce prismatic-shaped aggregates — tall, roughly hexagonal blocks, often bounded by polished surfaces.

Slaking

We have already noted that as the soil becomes wetter, so the stability of the aggregate decreases. One of the consequences of this reduced stability is the tendency for the aggregates to break down under the effect of slaking. By this process individual grains are detached and washed downwards through the soil. Detachment occurs mainly because as the soil is wetted air bubbles become trapped within the body of the soil. These escape by exploding, each explosion disrupting the bonds which hold the particles together. The more rapidly the soil is wetted, the greater is the tendency for air to be trapped. Slow wetting, on the other hand, such as that resulting from gentle drizzle, allows the air to escape before the surface of the clods is fully saturated. Consequently, slaking is most marked when a dry soil is quickly saturated by heavy rain or irrigation waters.

2.3.2 Sorting processes

Many of the weathering processes we have discussed also give rise to sorting of the soil particles. As a result of this sorting, the soil tends to display both a vertical and lateral pattern. Thus, in the vertical dimension, we can often recognise several horizons within a soil profile, each horizon having distinct physical properties. Laterally, we find that properties such as texture vary owing to the selective effects of geomorphological processes.

Freeze–thaw

Freeze–thaw activity leads to the separation of soil materials according to size. This sorting process operates over relatively small distances (a few metres) and produces patterned ground features such as **stone polygons** and **stone stripes**. The exact mechanism seems to vary to some extent, but in general these features result from the development of ice lenses within the soil. These lenses grow to a considerable size as water moves through the soil towards the frozen — and thus drier — areas. As a result, the soil is heaved upward. As the ice melts, fine particles preferentially settle, by falling through the crevices and pores between the larger stones. Over time, repetition of this process leads to complete sorting of the material and the stones accumulate in distinct polygonal zones. On slopes, the process is accompanied by solifluction (the downslope sludging of the soil saturated by melting of the ice) and thus the polygons become drawn out into less regular forms and, ultimately, into stone stripes.

Fossil features of this kind are frequently found in Britain, a legacy of the colder climates which existed during the Ice Age. Particularly fine examples are found in Breckland, in East Anglia, and in the river terraces of the Warwickshire Avon. Today, the activity is too subdued to cause similar features to be actively forming. Nevertheless, on a smaller scale the growth of needle ice can often be seen to lead to heaving of the soil and localised sorting of the soil particles.

Clay translocation

Slaking and wetting-and-drying act in combination to cause vertical sorting of the soil. While the effects of contraction upon drying and expansion on wetting lead to the development of deep, stable fissures in the soil, slaking detaches particles from the aggregates and allows them to be washed downwards through the soil. Clay-sized particles, in particular, are transported in this way, being moved from the upper horizons of the soil and deposited in the subsoil. This leads to a depletion of clay in the topsoil, and an accumulation below. In such circumstances, textural analyses show a zone of clay enrichment. It should, however, be mentioned that horizons of clay concentration may also occur through other processes, such as deposition of clayey layers in the original parent material.

The effects of **clay translocation** can be seen by the presence of thin, smooth coatings of clay over the surfaces of stones and aggregates, 'bridges' of clay across pore spaces, and veneers along worm and root channels. Often this clay is a different colour from that of the original, surrounding soil and thus is readily discernible. In other cases, however, it can only be seen by studying thin sections of the soil under a high-powered microscope.

Sorting on slopes

Both the processes we have discussed so far operate on a relatively localised scale; they lead to sorting of the soil *in situ*. Sorting on a larger scale results from erosion and deposition by water, wind and gravitational forces.

Possibly the most obvious example of sorting by winds is seen in dune areas, where soils show a marked textural gradation away from the shore-line. This is because winds transport sand landward, depositing the coarser particles near their source, and carrying the finer materials inland. Sorting by water and gravity, on the other hand, is widely found on slopes. The combined effects of solifluction, creep and wash result in the segregation of the soil according to particle size. In general, finer particles are moved further downslope than coarser particles. Thus, on the upper slope the soil is characteristically stony and coarse-textured, formed of residual materials left by these slope processes. At the slope foot, active erosion gives way to deposition and the soils are fine-textured. This relationship between slope and soil is known as a **catena** or **toposequence**.

These sorting processes have a number of other consequences. Where erosion is greatest, normally on the steep upper slopes, the soils tend to be thin and infertile. This reflects the fact that transport of the material down the slope proceeds more rapidly than weathering, and thus there is no net accumulation of weathered material. These have been called **weathering-limited slopes**. In some cases, in fact, erosion may be so active that the soil is completely removed (*Plate 3*), and the bare bedrock exposed; in less intensely weathered areas, the topsoil may be lost, revealing the subsoil and producing a truncated soil profile.

Plate 3. Slumping of soil on a steep roadside slope. Saturated by rainwater the soil slips downslope, leaving behind it the bare bedrock surface (Photo D. Briggs).

In contrast, on the lower slopes, accumulation of material derived from upslope leads to the development of deep soil profiles, often showing buried horizons resulting from the occasional incursion of debris washed from upslope. In addition, because little removal occurs from the footslope, the soil particles remain in position long enough to become intensively weathered. This, too, encourages the development of deep, clayey soils. A corollary of this is that the footslope soils tend to be poorly drained, for the high clay contents retard water movement through the profile. Thus, sorting on slopes leads not only to a textural and depth sequence but also to a hydrological sequence.

2.4 AGRICULTURAL IMPLICATIONS

2.4.1 Physical aspects of soil fertility

Soil texture has several important indirect effects on plant growth, through its control of factors such as structure, drainage and aeration. Clays also act as a major

TABLE 2.3 PERCENTAGE OF PORES OF DIFFERENT SIZE IN SOILS OF DIFFERENT TEXTURE

		Sandy loam	Clay
Macropores	60 μm	33%	10%
Mesopores	2–60 μm	33%	40%
Micropores	2 μm	33%	50%

store of plant nutrients, and therefore many aspects of soil fertility are ultimately influenced by texture.

One of the most significant indirect effects of soil texture is through its control of structure. Roots require space in which to grow, and this space is provided by the pores and fissures in the soil; in this way soil structure exerts a considerable influence on plant growth.

Plant roots extend into the soil pores in search of both water and nutrients. Since most roots are of the order of 1–5 mm in diamter, while the pores in the soil are rarely more than 1 mm in size (*Table 2.3*), roots can only grow by forcing their way into the soil. A critical factor, therefore, is the ability of the soil to give way in the face of the limited pressure which can be exerted by the tips of the growing roots. This in turn depends upon the **consistence** and bulk density of the soil.

In general, soils with a high bulk density have numerous small pore spaces, and thus only the tips of the relatively fine root hairs can gain any purchase there. In addition, owing to the close packing of the particles, there is little opportunity for the roots to push aside these particles and open up the pores. It has been shown that cereal roots can extend into a sandy soil only if the bulk density is less than 1.7–1.8 $\mathrm{g\,cm^{-3}}$, and into clayey soils only at bulk densities less than 1.5–1.6 $\mathrm{g\,cm^{-3}}$. Compacted soil horizons therefore present a formidable barrier to root growth. In the same way, high bulk densities limit seedling growth. If the soil above the seed is compacted the seedling is unable to force its way upward, and it fails to emerge from the soil. The effect is seen in the patchy growth of row crops such as cereals or beans.

As we have seen, the consistence of the soil tends to decrease as the moisture content increases; as the soil gets wetter root and seedling growth becomes easier. Thus, it has been found that whereas tap roots were unable to penetrate a soil with a bulk density of 1.75 when the moisture content was 5.5%, over 60% could penetrate when the moisture content was increased to 8%.

An interesting situation exists in many clayey soils where shrinkage and swelling occurs as the soil is alternately dried and wetted. When dry, large fissures open up and roots and seedlings can easily grow into them. When the soil is wetted, however, these fissures close up and the roots are pinched tightly by the peds. As a result, roots tend to be deformed and their ability to support the plant and supply nutrients is impaired.

2.4.2 Management of physical properties

In general, it is not feasible to alter significantly the texture of the soil, though, in a few instances, attempts are made to add material of a particular size grade. Thus, in some areas of very sandy soils, **marling** is practised; marl (a calcareous, silty clay) is added to the soil in order to increase its clay content.

In contrast, control of the soil structure is one of the main aims of soil management. Ploughing, harrowing, rolling — almost all the activities of tillage — are intended to break down the soil aggregates and produce a fine, even surface structure which will act as a good seedbed. In this state, the soil will allow the rapid and unhindered development of roots and the easy emergence of seedlings. Several additional benefits also accrue from tillage, however. **Ploughing**, for example, **helps to reduce erosion** by producing a rough surface and thus increasing the frictional resistance of the soil to both wind and water (*Plate 4*). **It also breaks up surface crusts**, caused by traffic or by the impact of raindrops, and thereby increases porosity. In this way it increases infiltration and reduces overland flow. In addition, **tillage removes weeds** and thus reduces competition from unwanted plants for the limited supplies of water and nutrients in the soil.

Not all the effects of tillage are beneficial, however. **Excess tillage may cause the aggregates to break down completely**; the individual particles thus formed will be susceptible to erosion (*Plate 5*). This is a danger particularly in silty or sandy soils, in which the natural stability of the structure is low. It is one factor which has contributed to the erosion of the sand and silt soils of the Vale of York, east Nottinghamshire, Lincolnshire and parts of East Anglia. As we will see later, tillage may also lead to the loss of organic matter and this too accentuates structural and erosion problems.

In addition, tillage often involves the use of heavy machinery which may damage the soil. Particularly when the soil is wet, the weight of large machines and, more important, the **shearing** caused by slippage of the wheels, may compact the soil. In

Figure 2.4 Smearing: the results of smearing by agricultural machinery on a soil developed from chalky boulder clay (from HMSO, 1970)

the same way, a **blunt plough share may cause smearing**; the soil beneath the share is re-oriented so that the clay particles lie parallel to the surface, as a thin veneer of densely packed material (*Figure 2.4*). Repeated ploughing to the same depth in this way can create an extremely compact **ploughpan** at the base of the topsoil. This hinders root development, reduces the rate of water percolation and acts as a barrier to air movement. As a result crop yields may be severely reduced. Again, this problem is more acute when the soil is wet, for in this state the structural stability and consistence is at a minimum. Consequently, it is important to cultivate the soil when it is in a suitable physical state; 'timeliness' of cultivation is therefore particularly important in heavy clay soils which are inherently susceptible to compaction and smearing.

Because of these difficulties involved in structural management of the soil, attempts have been made to reduce the amount of tillage necessary to produce a suitable physical environment for crop growth.

One such approach is **zero tillage**. In this case the soil is not cultivated at all prior to sowing, the seeds are simply sown in a narrow slit cut into the soil. After harvesting, the crop residues are left to decay in position. Clearly this minimises the danger of compaction or smearing by machinery, but it is not a technique which can be used in all circumstances. Without tillage, for example, the problem of weeds and pests sometimes becomes acute and it is necessary to use expensive pesticides to control them. In addition, the topsoil tends to become dense and relatively impervious without the action of ploughing to break up the surface. Several experiments have shown that as a result crop yields may be reduced under zero tillage. Nevertheless, in soils which are susceptible to structural damage it is obviously an approach which helps to conserve and protect the soil. In the long term, therefore, it may allow higher yields to be obtained.

2.5 CASE STUDIES

2.5.1 Introduction

As we have seen, textural analyses are particularly useful in studies of soil formation. They help to illustrate the processes of weathering and sorting which take part in

Plate 4. Ploughing lifts, breaks up and turns over the soil. The infiltration capacity of the soil is increased and erosion by water consequently reduced. The rough soil surface also reduces wind-speeds and prevents wind erosion (Photo D. Briggs).

Plate 5. Loss of organic matter from sandy soils allows wind erosion to occur. In this example in Lincolnshire a field sown to spring wheat is seriously eroded during a single blustery day; the sand blown from the field is banked up against an old hedge-line. In the background winds carry away more soil (Photo J. Owen).

soil development, and also give an indication of the rate' at which soil development occurs. Measurements of silt content have been used to identify the extent and importance of **loess** in soil formation. Textural analyses have also been used to study sorting processes within the soil, both by periglacial activities, such as freeze–thaw, and by the sorting action of clay translocation.

Soil texture is not important only as an index of soil formation. Indirectly, texture affects vegetation growth through its control of nutrient availability, drainage and organic activity. Various studies have therefore shown that soil texture is related to the yield of agricultural crops. An investigation in Oxfordshire by Clarke (1951), for example, indicated that almost 95% of the variation in the yield of wheat within a single field could be accounted for by differences in the texture, depth and drainage of the soil.

A more direct influence on crop yield is frequently exerted by soil structure. It has been shown that the fineness of the structure of the seedbed affects the emergence and therefore the yield of seedlings of oats (*Figure 2.5*), while a relationship

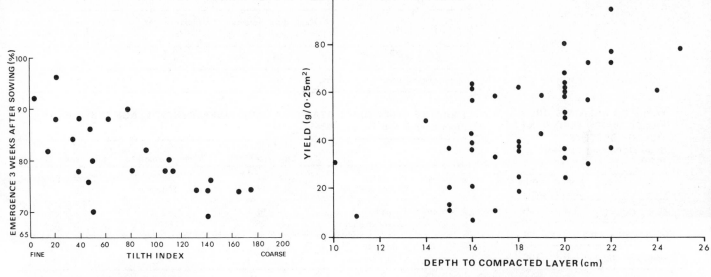

Figure 2.5 The relationship between soil structure and the emergence of seedlings of oats. The tilth index is a measure of the coarseness of the soil aggregates (from Thow, 1963)

Figure 2.6 The relationship between depth to compaction and barley yield: an example from a single field in Devon

is also found between the degree of surface compaction of the soil and the yield of cereals (*Table 2.4*). In addition, the structure of the subsurface soil influences crop yield through its control on root development and plant nutrition. It is possible to demonstrate that yields tend to increase as, for example, the depth to a compacted layer increases (*Figure 2.6*).

Several studies have also illustrated the effects of soil management upon the physical properties of the soil. The use of zero-tillage methods and traditional ploughing

TABLE 2.4 EFFECT OF COMPACTION UPON BARLEY YIELD

	Mean bulk density (g cm^{-3})	Mean height (cm)	Mean yield (g/0.25 m^2)
Compacted soil	1.84	21.3	137.5
Uncompacted soil	1.40	30.1	230.3

TABLE 2.6 RELATIVE CROP YIELDS UNDER DIFFERENT TILLAGE SYSTEMS
(from Davies and Cannell, 1975)

	Winter wheat	Spring barley
Direct drilling	94	90
Shallow ploughing	96	94
Deep ploughing	100	100

TABLE 2.5 STRUCTURAL STABILITY UNDER DIFFERENT TILLAGE SYSTEMS
(from Cannell and Finney, 1974)

		% Low stability	% Medium stability	% High stability
0–2.5 cm	Ploughed	46	47	7
	Direct drilled	20	44	36
2.5–5 cm	Ploughed	58	40	2
	Direct drilled	26	68	6
5–10 cm	Ploughed	30	60	10
	Direct drilled	25	58	17

methods have been compared, and it seems that while the structural stability of the soil is better under zero-tillage (*Table 2.5*), its bulk density is somewhat higher (*Figure 2.7*). Possibly because of this, yields of crops under zero-tillage commonly seem to be about 90% of those grown after traditional tillage (*Table 2.6*).

2.5.2 Aspect and soil formation
(Small, 1972)

The problem
One of the major factors affecting soil development is climate. Through its control on weathering processes and rates, on vegetation and on soil organisms, it influences the nature of the soil, both directly and indirectly.

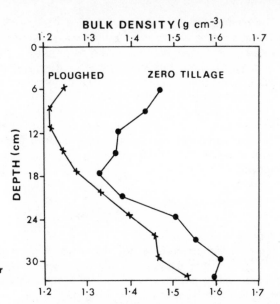

Figure 2.7 Bulk densities of soils under zero tillage and traditional ploughing (from Soane *et al.*, 1975)

It is difficult to study the effect of climate on a regional or national scale, for comparisons of soils in different climatic zones are often complicated by the unknown contributions from other factors. We cannot be certain, for example, that the history of soil management, or the time-scale of soil formation, are identical in our various sampling areas.

Fortunately, studies of **meso-climate** are much easier to design, and in many ways it is the meso-climate which has a dominant influence on soil formation. Aspect, in particular, provides a fundamental control upon the meso-climate, and thus investigations of soils in neighbouring areas of different aspect allow us to examine the role of meso-climatic factors in soil development. In the following example, the effect of meso-climate variations induced by aspect are considered in relation to both modern and buried soils in an area of western Wisconsin.

The method

The study area consisted of a loess-mantled valley running approximately from west to east. The loess had weathered to produce a soil about 0.6 m deep. This had subsequently been buried — apparently after settlement of the area about 100 years ago — by a thinner, modern soil, 0.3 m in depth.

On each side of the valley a grid 10 ft (∼3 m) by 80 ft (∼25 m) was pegged out. This was divided into 8 squares 10 ft × 10 ft. Within each square a sample point was selected by choosing co-ordinates from random number tables. At this point, the soil depth and slope aspect were determined in the field. In addition samples were collected from both the surface horizon of the modern soil and the upper horizon of the buried soil. These were used for textural analysis by the hydrometer method.

Results

The relationships between soil depth and slope position are shown in *Figure 2.8*. It can be seen that the soil initially decreases in depth in a downslope direction, but, on both valley sides, increases rapidly in the area within 50 ft (∼16 m) of the stream channel. On the north-facing slope this change in the trend of soil depth seems to coincide with a marked steepening of the slope angle; on the south-facing slope the change in gradient is more subtle. It is also apparent from this diagram that a considerable difference in soil depth occurs on the two slopes. Although both show the same pattern, the soil depth on the north-facing slope is consistently less than that on the south-facing slope.

Results of the textural analyses are given in *Table 2.7*. Comparison of the mean percentages of sand, silt and clay for the opposing slopes again suggests a significant difference in soil conditions. For example, the clay content on the south-facing slope is uniformly lower than that on the north-facing slope. It also appears that the clay content of the buried soil is uniformly greater than that of the modern soil.

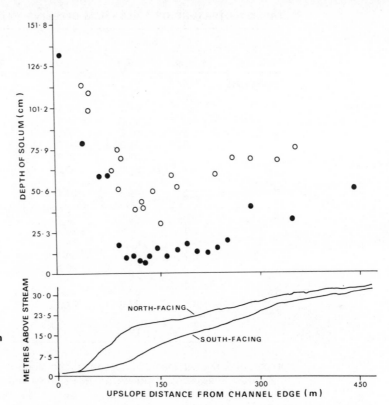

Figure 2.8 Variations in soil depth on slopes of opposing aspect (reproduced by permission from The Professional Geographer **of the Association of American Geographers, Volume 24, 1972, T.W. Small)**

Conclusions

These results can be interpreted in a number of ways. It would certainly seem that aspect has a major influence upon both the depth and the texture of the soil. The

TABLE 2.7 TEXTURE OF SOILS FROM OPPOSING VALLEY SLOPES

	Buried soil			Modern soil		
	% Sand	% Silt	% Clay	% Sand	% Silt	% Clay
South-facing	3.8	55.9	40.3	0.9	71.2	27.9
	3.9	56.7	39.4	0.8	73.4	25.8
	2.5	58.4	39.1	2.1	69.5	28.4
	4.0	60.8	35.2	4.0	66.7	29.3
	0.3	61.6	38.1	4.5	67.0	28.5
	1.0	58.9	40.1	0.3	68.8	31.2
	0.0	59.2	40.8	0.0	75.2	24.8
	1.7	58.3	40.0	2.7	72.7	24.5
$\overline{X}$	2.2	58.7	39.1	1.9	70.6	27.5
North-facing	2.0	64.3	33.7	0.0	74.7	25.3
	2.7	61.5	35.8	1.4	77.5	21.1
	3.0	62.9	34.1	0.0	77.6	22.4
	2.7	66.9	30.4	0.0	78.3	21.7
	4.2	68.9	26.9	1.6	76.3	22.1
	1.9	69.5	28.6	0.0	75.8	24.2
	8.2	67.4	24.4	0.1	76.2	23.7
	4.5	67.6	27.9	2.5	76.4	21.1
$\overline{X}$	3.7	66.1	30.2	0.7	76.6	22.7

deeper, and finer-grained soils occur on the slope of southern aspect, and this implies that weathering is more intense on this side of the valley. The fact that the buried soil is both deeper and more clayey than the modern soil may reflect the fact that the latter is relatively immature (less than 100 years old) and is still undergoing weathering.

It is also notable that there is a pronounced difference in slope form on either side of the valley. This, too, may derive from the differences in weathering intensity.

On the south-facing slope the profile is more subdued, and it seems that slope retreat has proceeded somewhat further than on the north-facing slope. Taken with the evidence from the soil analyses this strongly suggests that the south-facing slope experiences the more intense weathering conditions.

Comment

Although it seems valid to argue from this evidence that meso-climate in this area has significantly affected soil development, we must be cautious in how we interpret these results. In the first place no direct measures of climate were made; aspect was simply used as a **surrogate variable**. It may, of course, be reasonable to assume that aspect influences meso-climate, by its control over insolation and exposure for example, but without climatic data we cannot determine which climatic factors (if any) were responsible for the variation in soil conditions.

We must also be wary of interpreting results merely by looking at them. Such observations can be very misleading, and if we want to ensure that our interpretations are objective it is essential to use statistical methods to analyse the results. In fact, using statistical techniques, it is possible to show quite conclusively that the silt and clay contents of the soil on either side of the valley do differ significantly, and this gives support to our initial interpretations. Nevertheless, the point to make here is that inference by eye is liable to error; we will often find that two different people interpret results very differently in this way. Thus, as we mentioned in Chapter 1, it is generally advisable to base our final conclusions upon statistical analysis of the data.

The approach used in this study lends itself to many similar investigations. Meso-climate almost certainly affects the chemical, hydrological and biological properties of the soil, and thus the relationship between aspect and these variables can also be studied. In a similar way, climatic variables such as exposure and frost incidence may be investigated; in coastal areas exposure tends to decrease rapidly inland and this is reflected in the nature of weathering and soil formation; in areas of undulating relief, frost hollows can often be found in which physical processes of weathering are accentuated.

All these approaches suffer from a similar difficulty, however. Whenever attempts are made to compare soil development or soil conditions in different areas, there is always the danger that external factors, such as variations in parent material or land use, may interfere. Generally we are trying to isolate the influence of one or two variables, to simplify a complex environmental system. Yet, in most cases, we have no control over the system as a whole; all we can try to do is find situations in which we believe that interference from the myriad of other factors is minimal. Nevertheless, we must be aware of the possibility of this interference when we try to explain our results.

2.5.3 Land use and soil structure (Low, 1955)

The problem

If we compare the surface of a soil which has recently been ploughed up from pasture, with one which has been for many years under arable cultivation, we may be struck by the marked difference in appearance. In general, the old pasture soil seems fresher in appearance, more 'mellow'; the soil aggregates are perhaps coarser, and often considerably moister; the colour is generally darker. These differences are not entirely illusory. Soil which has been under grassland does indeed seem to have characteristic properties. In particular, the structure of the pasture soil seems to be improved. Several experiments have been conducted to illustrate this effect. The example we will consider here looks at the changes in structural stability of pastures over time, and compares them with variations found in soils under continuous arable cultivation.

The method

Twenty-two farms were selected in England and Wales. On each farm an old pasture field was compared with a nearby field recently returned to pasture after a long period of arable cultivation. In each field a sampling area 20 yd X 20 yd (18.5 m X 18.5 m) was defined, and samples collected in the autumn and spring. In several fields sampling was continued for a period of nine years, so that a comparison of changes in the condition of the old and new grassland could be made.

The soil samples were taken to the laboratory, oven dried, then sieved through a

½ in (1.2 cm) mesh sieve. The fine fraction (less than 1.2 cm) was then used to determine the aggregate stability of the soil. In addition, the moisture content of the soil at the time of sampling was determined by the gravimetric method (see Section 3.2).

Results

A selection of results from the analysis are presented in *Figures 2.9* and *2.10*. A major feature of the results is the marked difference between the stability of the

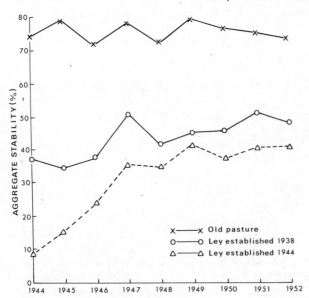

Figure 2.9 Annual changes in the aggregate stability of soils under different cropping systems (from Low, 1955)

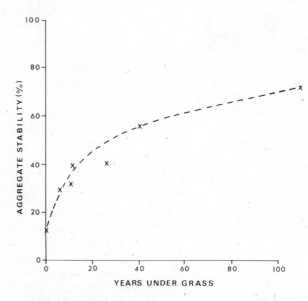

Figure 2.10 Long-term changes in aggregate stability of soils under grass (from Low, 1955)

old and new pastures. Whereas the old pasture soils had an average structural stability of 70–80%, the land newly returned to grass averaged only 5–10%. It is also clear that even after return to grass the soil structure does not immediately regain its optimum stability. In the fields which were monitored for a period of nine years, the structural stability increased markedly, but only reached a value of 40%. One field which had been under grassland for 15 years had only a marginally higher structural stability (43%).

It seems, therefore, that although the improvement in the stability of the soil structure under pasture is relatively rapid at first, the rate of improvement declines considerably after several years. Subsequently, the increase in stability is extremely slow and it may take up to 100 years for the stability to reach an optimum level.

Conclusions

The results obtained in this experiment allow three general conclusions to be reached.

1. A marked improvement in the structural stability of the soil occurs when it is put down to grass.
2. The improvement is rapid during the first three to four years, but subsequently takes place at a slow rate.
3. Between 40 and 100 years may be needed for equilibrium to be reached and an optimum stability (*c*. 70–80%) to be achieved.

Comment

While the time necessary for repeated annual measurements is obviously unavailable for most of us, studies similar to this can be carried out. Comparisons of nearby fields which have been under grass for different periods of time (as in the example quoted in *Figure 2.10*) are quite simple to design. It is also possible, of course, to invert this approach and study the effects of arable cultivation on the deterioration of soil structure, by comparing fields which have been under cereals for varying lengths of time. It is an interesting question whether it takes as long for the structure to be destroyed as it does to restore it.

Numerous other investigations are possible based upon the same principles. Measurements of other physical properties such as bulk density and porosity may be made. Changes in these properties under different types of land use can be studied; the effects of afforestation or of forest clearance, the effects of reclaiming soil from the sea or industrial wasteland, the effects of growing and, in particular, harvesting root crops, can all be studied. In every case it is necessary simply to obtain areas representing the soil before and after treatment. In many cases these areas can be found adjacent to each other, and this has the advantage of reducing any differences which may be due to other factors (such as inherent differences in soil type). In other cases analysis can be carried out on a single plot of land, immediately prior to treatment and again soon afterwards.

CHAPTER 3 HYDROLOGICAL PROPERTIES

3.1 INTRODUCTION

3.1.1 Classification of soil water

Water plays a major part in almost all the physical, chemical and biological processes in the soil. It is involved in most forms of mechanical weathering, is fundamental to all aspects of chemical weathering, redistributes material throughout the soil profile, carries away both soil particles and solutes, and transports nutrients to plants. Moreover, the soil water is immensely variable, both in quantity and quality, over both time and space.

Water enters the soil mainly from rainfall. It **infiltrates** the soil by moving into fissures and pores in the soil surface. The rate of infiltration depends on several factors, including the intensity of rainfall and the amount and state of pores and fissures in the soil. When the soil is already saturated by previous rainfall, for example, infiltration is reduced. Infiltration is also slow when the soil surface is compact and dense; when the pores are small and few in number.

If rain-water cannot infiltrate the soil it tends to run off over the surface. Although in rough terrain, or on low angle slopes, its surface movement is slow and water may be stored in depressions until it can infiltrate or evaporate, on steeply sloping ground the rate of run-off may be rapid and the water is able to pick up and erode soil particles. In this way gullies may develop and considerable losses of soil may occur.

The water which enters the soil is affected by two main forces: **gravity** and **matric** forces. **Gravity is responsible for the movement of water out of the soil by drainage. Matric forces, on the other hand, are responsible for the retention of water** in the soil since they lead to the attraction of water molecules toward the soil particles.

The operation of these forces is greatly dependent upon the size of the pore spaces in the soil. In general, gravity is only effective in pores with a diameter of more

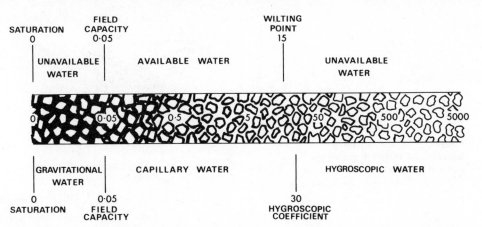

Figure 3.1 The classification of soil water according to biological (upper) and physical (lower) criteria. The figures represent soil moisture tensions measured in bars

than 0.06 mm; consequently, free drainage of water only occurs through the macropores. Within the micropores, the water is held more or less firmly by the forces of retention. This means that the soil moisture can be rather arbitrarily divided into several categories (*Figure 3.1*). That in the larger pores, and subject to gravitational forces is known as **gravitational water**. That in the micropores is called **capillary water**. The capillary water is able to move through the soil only very slowly, and, in theory, cannot readily drain out of the soil profile.

In addition, two further categories are sometimes identified. Some of the water within the micropores is held so tightly to the soil particles that it is almost immobile; it is in fact adsorbed on to the particles by an electrochemical bond. This is called **hygroscopic water**. It can be removed only by heating or prolonged drying. A small proportion of water is also bound up within the structure of the soil particles; for example, within the crystal lattice of the clay minerals. This is known as **structural water**, and is only released by destruction of the clay particles. The hygroscopic and structural water is therefore unimportant in terms of processes of water movement, but it does become significant when we try to measure the moisture content of the soil.

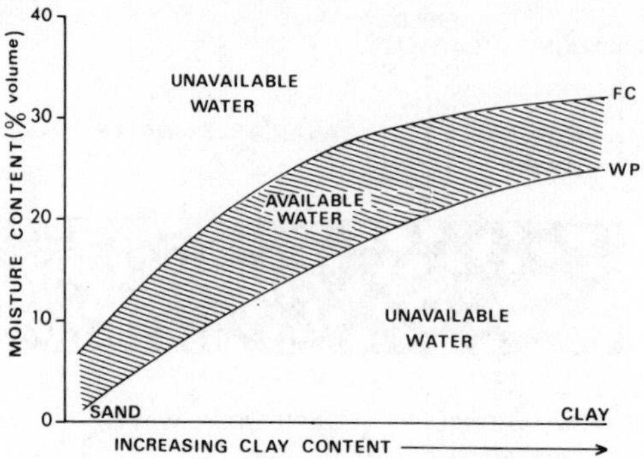

Figure 3.2 The relationship between soil moisture availability and soil texture

The boundaries between these various types of soil water are obviously not clear cut; they merely represent transitional phases where the character of water movement and retention gradually change. One boundary is of particular significance, however; namely, that between the gravitational and capillary water. This is frequently referred to as the **field capacity**, for it delimits the maximum amount of water which can be retained in the soil for any length of time.

This classification is based upon the physical processes of movement and retention of water, but moisture is also important in terms of plant growth. Thus a biological classification of soil water is frequently used (*Figures 3.1* and *3.2*). This divides the soil moisture into two broad classes: that which is available to plants and that which is unavailable. The **unavailable water** includes two separate components. Gravitational water is unavailable because it drains too rapidly from the soil after rainfall (often within 2 days). The structural, hygroscopic and part of the capillary water is also unavailable because it is held too tightly within the soil to be extracted by plants. Consequently the available moisture is composed of a relatively narrow range of water, which includes most — but not all — of the capillary moisture. It is bounded at the wet end of the range by the field capacity, while at the dry end it is delimited by what is known as the **wilting point** — the stage at which plants show evidence of permanent wilting owing to lack of moisture.

3.1.2 Soil moisture movement and measurement

We have already mentioned that two main forces are involved in the movement and retention of water in the soil. To understand these more fully it is necessary to delve a little deeper into the physics of water retention.

Water in the soil has what is known as **free energy**. Simply, this means that it has a potential — a tendency — for movement and change. The force of gravity encourages this movement, and thus increases the free energy of the soil water. In contrast, matric forces, by retaining moisture in the soil, prevent movement and reduce its free energy. In view of this, while gravity is referred to as a positive force, matric forces are considered to be negative forces or, more commonly, **tensions**. **Water is therefore held in the soil by tension**; to measure the strength of retention we must measure the soil moisture tension. The scale of measurement which we normally use is the **bar** or **atmosphere** scale.

Matric forces consist of two related processes. Water molecules are attracted to the soil particles by **adhesion**; in turn, the water molecules attract others by **cohesion**. In this way the water forms thin films around the soil particles. The forces of adhesion and cohesion are at a maximum in immediate contact with the soil particles. With increasing distance the strength of these forces decreases and, at a distance of about 0.06 mm from the particle surface they are negligible (*Figure 3.3*). Hence, **water molecules further than 0.06 mm from the soil particles are not affected by matric forces** — they are free to move through the soil under the influence of gravity. It is for this reason that, as we stated earlier, the character of water movement is related to the size of the pore spaces. In very small pores (the micropores) all the water is sufficiently close to the soil particles to be affected by matric forces. In the macropores, however, only the water lining the pore sides is retained in this way.

Although adhesion and cohesion are largely responsible for retention of water in the soil, they also allow moisture to move through the soil, independent of gravity. This is a result of capillarity. In simple terms, the attraction of the water molecules both to the soil solids and to each other causes the water to move from wet areas, where the matric forces are weak, towards drier areas where the water films are thinner and the matric forces stronger. In other words, the moisture moves down

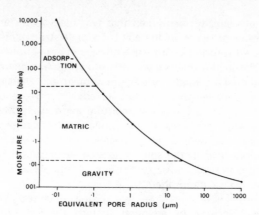

Figure 3.3 The relationship between soil moisture tension and pore radius

a gradient, from weak tensions to strong tensions. This movement may be in any direction: upwards, downwards or laterally.

In order to explain more clearly the effects of soil moisture tension let us consider the events following a heavy storm. Initially, water infiltrates the soil, filling the soil pores. **At the point where the soil is fully saturated, the soil moisture tension approaches zero**. Thus, an extremely small force would be necessary to remove water. Because of this, gravity is sufficient to cause drainage of part of the moisture content.

Gravitational drainage continues for about 2 days, until the water films in the soil pores are about 0.06 mm or less in thickness. At this stage, the macropores are almost emptied of water, but the micropores remain saturated. **The soil moisture tension is about 0.05 bar***; in other words, matric forces are exerting a tension of about 0.05 bar on the outside of the water films. From this point water loss by drainage is negligible; most of the water is removed by evaporation and plant uptake. In addition, a slow redistribution of water occurs through capillary movement.

*1 bar = 10^5 Pa

Ultimately, as the water films get thinner, and the soil moisture tension rises to about **15 bar**, plants start to show evidence of wilting. This is partly because capillary movement is by then so slow that it cannot meet the needs of the plant for water. Wilting point is therefore reached. Further water loss may occur through evaporation, though this degree of drying is rare in all but the most arid environments.

It is clear that in most cases the soil does not have a chance to reach wilting point, at least in normal British conditions, for renewed rainfall soon replenishes the water lost by drainage and evapo-transpiration. It is also clear that although drainage losses mainly occur when the soil is close to saturation, downward movement of water through the soil continues for a long period after field capacity has been reached.

3.2 TECHNIQUES

3.2.1 Soil moisture content

Introduction

It follows from what we have said that the amount of water in the soil varies markedly over time. During rainfall, for example, the moisture content increases, particularly near the surface. Afterwards, drainage, plant uptake and evaporation lead to a slow reduction in moisture content. Clearly, one of the most useful hydrological techniques is the measurement of soil moisture content. Through this we can monitor temporal changes in soil moisture, as well as spatial variations relating to differences in soil type and moisture retention. It also enables us to study a single profile and measure the downward movement of water through the soil; in general the water moves as a fairly distinct zone, bounded by a **wetting front**. Once the gravitational water has been lost, the capillary water behind this wetting front moves slowly towards drier areas of the soil. It may, in fact, take many days for this wetting front to penetrate the whole soil, for capillary movement is extremely slow.

In addition to measurements of hydrological processes, however, the analysis of soil water content forms a fundamental part of many other analyses. Thus we may need to determine the moisture content of the soil when we are carrying out chemical analyses of samples. We also need to measure the moisture content of the soil when determining its field capacity.

Measurement

Equipment: Oven set at 105°C
Balance
Crucible or beaker
Dessicator containing calcium chloride or silica gel

Sample: *c*. 50 g fresh soil, or a core sample of known volume.

Procedure: Weigh the moist soil; record the weight as W_1.

Place the soil in the beaker and dry in the oven for 12 hours.

Remove and cool in a dessicator.

Weigh the dry soil and record the weight as W_2.

Calculation: The moisture content of the soil is given by the following formulae:

$$M_w \text{ (\% wet weight)} = \left(\frac{W_1 - W_2}{W_1} \right) . 100$$

$$M_d \text{ (\% dry weight)} = \left(\frac{W_1 - W_2}{W_2} \right) . 100$$

$$M_v \text{ (\% volume)} = \left(\frac{W_1 - W_2}{V} \right) . 100$$

Estimation of the total moisture content of the soil is essentially very simple. A soil sample is collected and weighed in its field state, then thoroughly dried in an oven and reweighed. The weight loss gives a measure of the moisture content.

Despite the simplicity, several practical and conceptual considerations must be borne in mind. First, it is clearly essential to store the sample so that no water is lost between collection in the field and treatment in the laboratory. The sample should therefore be sealed in an airtight polythene bag, or well wrapped in aluminium foil, immediately it is removed from the soil. The method of drying is also important. If the soil is dried at too low a temperature, or for too short a time, all the moisture will not be removed; on the other hand, if the temperature is too high or the drying time too long, there is a danger of driving off organic compounds, structural water or some volatile inorganic compounds. Normally, the soil is dried for 12–16 h at 105 °C. For detailed work, however, the sample is returned to the oven for periods of 4–8 h and reweighed. This procedure is repeated until the change in weight is less than 1%.

A final consideration is the method of expression. The moisture content can be expressed as a percentage of the wet weight, the dry weight or volume of the soil. If it is based upon the wet weight the possible range of moisture contents is from 0 to 100%, but the relationship between percentage moisture content and volume of water in the soil is not constant. Thus, a soil containing 20% moisture is not necessarily twice as wet as one containing 10%. Use of the dry weight avoids this problem, but it may result in values of over 100%. In peats, for example, values of 200% may frequently be found; intuitively, this may be difficult to comprehend. Consequently, the most suitable means of expression is as a percentage of the soil volume. Since the specific gravity of water is about 1.0 g cm^{-3}, the weight loss (in grams) gives the volume (in cm^3) of the water in the sample. This method, however, requires the collection of a soil sample of known volume. This is most easily done with a core tube made, for example, from a length of metal piping or an old tin can.

3.2.2 Infiltration capacity

Introduction
Infiltration capacity refers to the rate at which water can enter the soil. It depends upon numerous factors, but in particular is affected by the physical characteristics of structure and texture. However, the antecedent rainfall conditions may also influence infiltration capacity and therefore, in order to standardise measurement,

infiltration is normally assessed under defined moisture conditions — for example, in a dry soil or in a fully saturated one.

The importance of the infiltration capacity has already been noted. It controls the rate at which rain-water can be taken into the soil and thus influences the point at which overland flow occurs. In soils with a low infiltration capacity, overland flow occurs at low rainfall intensities and thus the dangers of erosion are great (*see Plate 1*). Soils with higher infiltration capacities are able to absorb greater rates of rainfall input and are therefore less susceptible to erosion. It must be stressed at this point that **the critical factor is the condition of the soil surface**, not the subsurface layers. As we will see later in this chapter, infiltration, at least in an unsaturated soil, does not depend upon the rate at which water moves **through** the profile, but upon the rate at which it can actually pass across the 'interface' of the soil and atmosphere. Thus, the effect of surface crusts, compaction and tillage are very important.

Measurement

Equipment: Cylinder infiltrometer (metal cylinder *c*. 20 cm diameter, 30 cm deep; 5 l water bottle and constant head apparatus)

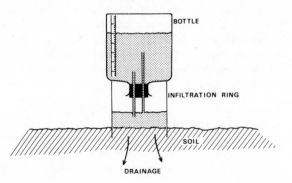

Figure 3.4 Use of the cylinder infiltrometer

Procedure: Calibrate the capacity of the inverted water bottle.

Fill the bottle with water; push the cylinder firmly into the soil (to a depth of at least 5 cm) and set up the equipment as in *Figure 3.4.* Allow the cylinder to fill with water to the level of the constant head apparatus, then time the rate of water loss from the bottle; take readings after 30 s, 1 min, 2 min, 5 min and 10 min, then as necessary.

Calculation: Plot the curve of cumulative water loss (infiltration) against time. Calculate the rate of infiltration into either the dry or wet soil (see text) as ●

$$I_c \ (\text{cm min}^{-1}) = \frac{\Delta Q}{A\Delta t}$$

in which ΔQ is the change in the volume of water in the bottle during time Δt, and A is the cross-sectional area of the cylinder (πr^2).

The simplest method of measuring the infiltration capacity of the soil is with a cylinder infiltrometer. This consists of a metal or strong plastic cylinder and a water bottle with a constant head apparatus. The cylinder is pushed into the soil, the bottle inverted above it and the cylinder allowed to fill with water. The rate of water loss from the bottle is then measured to give an estimate of the infiltration capacity.

In general, this method produces rather high rates of infiltration initially, but as measurement is prolonged so the infiltration capacity falls. The reason is that, as we have seen, antecedent moisture conditions — whether the soil is wet or dry — affect the rate of infiltration. When the soil is dry its infiltration capacity is at a maximum, but as it approaches saturation, then the rate of percolation starts to exert a controlling influence on infiltration. Thus, the values of infiltration capacity show a roughly parabolic trend over time.

This raises some problems of interpretation, for it is clearly impossible to define a single value of infiltration capacity which is characteristic of a single soil. Instead, it seems more appropriate to define two such measures; the infiltration capacity of the dry soil and that of the saturated soil. The former value is given by the slope of the first part of the curve and is often called the **instantaneous infiltration capacity**; the latter is given by the gentler, second limb and may be known as the **saturated infiltration capacity**.

A second factor must be borne in mind when interpreting these curves. This is that measurement is based upon calculating the rate of infiltration of water from a flooded surface. In reality, however, infiltration is occurring through the entry of rainfall directly into the soil. In this situation, the infiltration capacity tends to be rather lower, and to decline somewhat over time owing to the compaction and structural damage caused by the pounding of the raindrops. This effect is most marked in intense storms, and thus rainfall intensity affects the nature of the infiltration curve. In other words, the cylinder infiltrometer gives only a general indication of infiltration capacity; it does not measure infiltration under real-world conditions.

3.2.3 Field capacity

Introduction

A second permanent hydrological property of the soil is field capacity. This measures the amount of water which is retained in the soil after all tne freely draining gravitational water has been removed. It thus supplies an estimate of the ability of the soil to lose or retain water.

Various factors affect the field capacity. The most direct influences are exerted by the structural properties of the soil; in particular the pore size distribution. In soils which are dominated by large pores (e.g. sands) gravitational water accounts for the vast majority of the water in the soil. After a storm this is rapidly lost and thus the proportion held at field capacity is small. Conversely, in clayey soils in which the majority of the pores are small, matric forces affect almost all the moisture entering the soil. As a result little water can drain away and the moisture content at field capacity is high.

As this implies, soil texture has an indirect effect upon field capacity, through its control of structure and porosity. In addition, several other factors operate in an indirect way. Land use and agricultural activities, for example, affect the porosity of the soil; compaction by machinery and animals tends to reduce the proportion of large pores in the soil and increase the amount of water held at field capacity. Ploughing, on the other hand, breaks up the soil aggregates and increases the number of macropores; as a result, the water drains more easily.

The field capacity of any soil horizon does not depend only upon the properties of that horizon. It is also affected by the nature of the underlying layers. Thus, the drainage and percolation rates of the subsoil have an important influence upon retention in the topsoil. The presence of an impervious layer, such as a clayey horizon, a compacted layer, or an ironpan, may therefore impede drainage and increase the field capacity of the overlying soil.

Measurement of field capacity has a number of uses. It provides an estimate of the amount of water which may be retained in the soil after rainfall, and thus gives a rough indication of the potential supply of water to plants. In addition, measurement of field capacity helps define the suitability of the soil for cultivation. Generally, it is believed that **the soil should not be tilled until it is below field capacity**; that is, until free drainage has ceased and some additional water has been lost by evaporation or plant uptake. Above this level, the soil is too moist for cultivation and is liable to smearing and structural damage.

Measurement

Equipment: 3 core tubes
Balance
Oven set at 105°C
Dessicator containing calcium chloride or silica gel
Fine cloth and elastic band
Bowl or basin
Sand tray filled with 2–3 cm of dry, fine sand

Sample: Core samples, *c*. 2–3 cm in height

Procedure: Collect undisturbed core samples; calculate the volume of the cores ($V = \pi r^2 h$).

Cover one end of the cores with cloth, held in place with elastic bands; cover the other end with tin foil to prevent evaporation.

Place the samples in the bowl of water, deep enough to come within 1 cm of the top of the soil; leave for 8–12 h.

Remove the sample from the water; stand the core on the sand tray and leave for 48 h.

Remove the soil from the core; weigh the soil; record the weight as W_1.

Place the soil in the oven and dry at 105 °C for 24 h; cool in a dessicator and weigh; record the weight as W_2.

Calculation: The field capacity is the moisture content of the soil after drainage; it is given by

$$F_c \text{ (\% dry weight)} = \left(\frac{W_1 - W_2}{W_2} \right) . 100$$

$$F_c \text{ (\% wet weight)} = \left(\frac{W_1 - W_2}{W_1} \right) . 100$$

$$F_c \text{ (\% volume)} = \left(\frac{W_1 - W_2}{V} \right) . 100$$

N.B. It is advisable to use three replicate samples and calculate the mean of the three results.

The simplest approach to measuring the field capacity is to calculate the moisture content of the soil after it has been allowed to drain from saturation for two days. This is the method described above; it simply involves taking an undisturbed core sample, saturating it by standing the core in water for several hours, and then allowing it to drain for 48 h. Drainage is best carried out on a layer of dry sand, but can also be achieved by suspending the soil core in a retort stand. After two days the soil is weighed, dried in an oven and reweighed, the weight loss providing a measure of moisture content at field capacity.

This approach is simple, but its theoretical foundations are weak. The time taken for gravitational water to drain away varies considerably according to the texture and structure of the soil. In sandy soils, as we have seen, water is lost rapidly after rainfall, and a distinct point is reached at which drainage ceases. This occurs after about two days and gives a reasonably clear definition of field capacity. In heavy soils, however, drainage is slow and may continue for many days or even weeks after rainfall. In this case an exact definition of field capacity is not possible.

One way of avoiding this problem is to measure the moisture content not after a predetermined period of drainage but at a specific tension. As we have already noted, field capacity generally occurs at a soil moisture tension of about 0.05 bar. By measuring the quantity of water held in the soil at this stage, we have a means of assessing the field capacity (or, more strictly, the 0.05 bar percentage, as it is called).

To measure this parameter involves applying a force of 0.05 bar to the soil so that all the water held by tensions lower than this is withdrawn. A soil core is placed on a filter paper in a Buchner funnel and the surrounding space filled with sealing wax. The funnel can then be placed in a vacuum flask and attached to a vacuum pump (*Figure 3.5*). A vacuum of 0.05 bar is then developed in the vacuum flask, and the system allowed to come into equilibrium; at this point all the water held at tensions of less than 0.05 bar will have drained from the soil core. The moisture content can then be determined by oven drying.

3.2.4 Saturated hydraulic conductivity

Introduction

As we saw early in this chapter, two types of water movement through the soil take place. When the soil is saturated, movement is mainly by drainage under the influence of gravity. When the moisture content falls below field capacity, however,

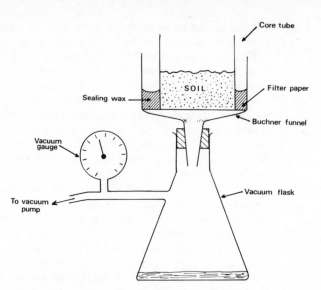

Figure 3.5 Determination of field
capacity by the vacuum method

capillary movement dominates, the water moving independently of gravity from
moist areas (low tension) towards dry areas (high tension).

It is difficult to measure the rate of capillary movement, but various estimates
have suggested that in a loamy soil a rate of about 0.0002–0.002 cm/day is attained.
The rate of movement varies, however, according to the soil moisture tension. It
increases markedly as field capacity is approached, and falls as the moisture content
is reduced. By the time wilting point is reached the rate of capillary movement is
practically zero.

By comparison, gravitational flow is extremely rapid. Water in a completely saturated sandy loam soil may move downwards at a rate of 100 cm/day or more. In
heavier soils, the rate of movement is considerably slower; in a clay, for example,
flow rarely exceeds 10 cm/day. This indicates that one of the main factors controlling the rate of saturated flow is soil texture. As we have seen before, this

operates indirectly through its influence on the structure and, more specifically, pore size distribution of the soil. Water moves slowly through small pores, relatively rapidly through large ones.

This relationship between pore size and rate of saturated flow can be more clearly defined. **In the words of Poiseuille's law, the rate of flow is proportional to the fourth power of the radius of the pores**. Thus, halving the diameter of the soil pores reduces the rate of flow sixteenfold (2^4). It follows from this that the rate of movement rises very rapidly as the size of the soil pores increases. Because of this, percolation is enhanced by any factor which encourages the development and maintenance of large soil pores. Organic matter, plant roots, earthworms, efficient (but not over-zealous) cultivation, the installation of tile or mole drains (which are really ultra-large soil pores) all increase the rate of saturated flow.

Another factor affecting the rate of saturated flow is the **hydraulic gradient**; the difference in height between the level of percolating water and the level to which it is draining — normally the water table at the base of, or below, the soil. Because of this, water movement in a saturated soil tends to be greatest close to the surface and to decline as the water table is reached. In addition, the hydraulic gradient provides a force for lateral flow in a saturated soil.

The rate of lateral movement is considerably slower than vertical movement, however, for it is not aided by gravity. Nevertheless, it is an important aspect of water movement for it helps supply water to tile drains and to natural pipes in the soil.

The rates of both vertical and lateral movement of water through the soil are important for two reasons. Firstly, they control the rate at which excess water is lost from the soil, and at which the soil returns to field capacity. This in turn determines how long after rain the soil remains in a wet and unworkable condition; it also affects the supply of oxygen to plant roots. Secondly, where percolation is slow it may lead to prolonged saturation of the upper layer of the soil. This may in turn reduce the rate of infiltration and cause overland flow to occur during rainfall. The result may then be soil erosion (*see Plate 1*).

Measurement

Equipment: Core tube
Retort stand and clamps
Cloth and elastic band
500 ml flask with constant head device
250 ml flask
Funnel
Measuring cylinder

Sample: Core sample

Procedure: Measure the radius of the core sample and calculate its cross sectional area (πr^2).

Cover the lower end of the core with cloth, held in position with an elastic band.

Place the core in a basin of water (the water level should be 1–2 cm below the top of the soil); leave overnight or until the soil is saturated.

Set up the equipment as in *Figure 3.6*; measure the depth of soil in the core (h) and the hydraulic head (ΔH).

Allow the water to drain through the soil core for 10 min–1 h; remove the lower flask and measure the volume of water collected.

Repeat the process at intervals of up to 24 h, until about 6 readings are obtained.

Calculation: Calculate the hydraulic conductivity (K) for the soil, thus:

$$K \text{ (cm min}^{-1}) = \frac{Q}{At} \cdot \frac{h}{\Delta H}$$

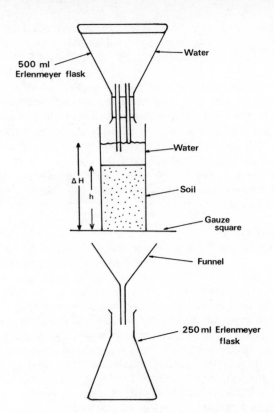

Figure 3.6 The determination of
the saturated hydraulic conductivity

where Q is the volume of water (in cm^3) collected during time t (minutes); A is the cross-sectional area of the core (cm), h the depth of soil (cm) and ΔH the hydraulic head (cm).

N.B. It is advisable to calculate a separate value of K for each reading; any 'rogue' values may then be discarded before calculating an average hydraulic conductivity.

Although use of this method is simple, several problems do arise. In particular, difficulty may be met in obtaining an undisturbed core sample; compaction or cracking of the soil during collection will give false measures of hydraulic conductivity. In addition, it is sometimes difficult to obtain a complete seal around the edge of the soil, and leakage occurs down the inside of the core tube. This is easily overcome by greasing the inside of the core tube with silicone wax or petroleum jelly prior to sampling. Even so if accurate results are required it is advisable to carry out three or more replicate analyses and take the average reading (omitting any that are obviously in error).

3.2.5 Water table level

Introduction

In freely drained soils, water drains rapidly after rainfall and only rarely accumulates within the soil profile. Where drainage is impeded, however, a zone of saturation may develop and this often persists for much of the year. The level of the saturated layer — the **water table** — varies according to the balance of inputs from rainfall and the losses through drainage and evapo-transpiration.

Two types of drainage impedance can generally be recognised. In some cases, poor drainage is a result of impervious layers below the soil, or of the position of the soil, in a low-lying area from which water cannot escape (e.g. in estuaries or inland basins). In this situation, the impedance is said to be due to **ground-water** conditions and the soils which develop are known as ground-water gleys. This contrasts with soils in which drainage is restricted by impervious layers within the soil profile, such as clayey horizons, ironpans or compacted layers. These are known as **surface-water gleys** (or, in more recent terminology, **stagnogleys**).

As we will see later, periods of saturation leave their mark on the soil by causing reduction of iron and aluminium oxides. This produces grey or green colours in the soil. Conversely, when the soil dries out, oxidation takes place creating red and yellow colours. Where alternating periods of saturation and dry, well-aerated conditions occur, the soil is therefore characterised by patches of both reduced and oxidised material. The resulting mixture of grey and yellow colours is known as **mottling**. Soils in which reduction has occurred are said to be **gleyed**.

In ground-water gleys, the lower horizon of the soil can be almost permanently saturated, and thus reduction is the normal process. The material is therefore intensely gleyed and grey colours predominate. In the upper layers, however, the soil may dry out during the summer, and thus oxidation is able to occur, providing a mottled horizon.

By contrast, surface-water gleys experience most prolonged saturation at or near the surface, and thus the layer of most intense gleying is often perched within the soil. Below, the soil may be relatively brightly coloured. As this implies, soil colours provide a means of inferring the drainage status of the soil, and for field work they help to locate the main zones of permanent saturation, temporary saturation and permanent aeration in the soil. In addition, it is possible to use the soil colours to give a classification of soil drainage (*Table 3.1*).

Although soil colour gives us a clue to the general status of soil drainage, for a variety of reasons it is only a partial picture. This is largely because soil colours are influenced by factors other than water conditions. The natural colour of the parent material, in particular, has a major effect; soils produced from red sandstones and clays, such as the Permian and Triassic rocks, are characteristically red-coloured themselves, even in relatively poorly aerated conditions.

In addition, the colours produced by oxidation and reduction survive for very variable periods, depending upon the local soil conditions. In some cases colours may develop and disappear within a single season; in other cases they may persist for centuries after the processes which created them have ceased to function. It is often difficult, therefore, to know whether the colours which we see today are inherited from a previous phase of soil formation, or whether they were a product of a single, possibly exceptional season. Consequently, it is far more informative to measure the extent of saturation in the soil. Even though measurement may only be possible for a relatively short period (e.g. one season, or one year) the

TABLE 3.1 CLASSIFICATION OF SOIL DRAINAGE FROM GLEY MORPHOLOGY

Drainage class	Drainage conditions	Gley morphology
Free	No waterlogging above 80 cm except during first four days after heavy rainfall	No mottling in upper 80 cm
Moderate	Waterlogging above 60 cm for less than one month	Distinct mottling below 60 cm
Imperfect	Waterlogging above 60 cm for more than 1 month continuously	Distinct mottling below 60 cm; faint mottling 30–60 cm
Poor	Waterlogging for long periods during the summer	Distinct mottling below 30 cm
Very poor	Almost continuous waterlogging	Grey colours in and below the topsoil

results can be related to rainfall conditions, and an idea obtained of the general hydrological properties of the soil. Moreover, as we will see, it is far easier to distinguish between surface- and ground-water impedance by actually monitoring the level of the water table in the soil. We will also see that this is of considerable importance when trying to devise a method of artificially draining the soil.

Measurement

Equipment: Large-bore auger (e.g. Dutch or Post-Hole auger)
Porous plastic or tile drain (optional)
Tin plate or small tin can
1 m steel tape or rule

Procedure: Using the auger, drill a hole to the desired depth; if necessary line the top of the hole with a short (30 cm) length of porous drain pipe.

Cover the top of the hole with a steel plate or small tin can to prevent contamination.

At intervals of 1 day–1 week, measure the depth to the water table in the hole; measure the total rainfall at the same time.

In clayey soils, variations in the level of the water table may be very easily measured. An auger hole is drilled, using a Dutch or Post-Hole auger (to give a hole *c*. 10 cm in diameter) and the hole is allowed to fill up with water. The level which the water reaches in the hole will, for all intents and purposes, be equal to the level of the soil water table. By measuring the depth from the surface to the water at intervals of about one week (more frequently if detailed results are required), the pattern of variation can be ascertained.

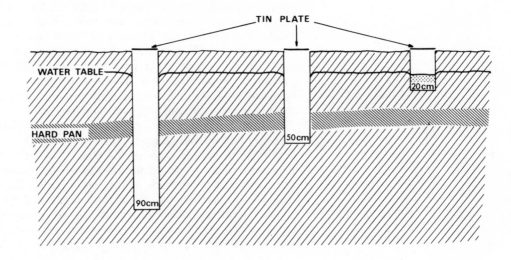

Figure 3.7 Measurement of the water table level by the dip-well method

If the amount of rainfall is also measured, the relationship between precipitation and water-table level can be determined. It is also interesting to continue measurements throughout the period from early autumn to mid-spring. In general, a marked rise in the water level will be seen at the end of the autumn and a sharp fall at the beginning of spring. This is due not so much to changes in the amount of precipitation, as to changes in the rate of evaporation and transpiration. During the autumn, temperatures fall and evaporation is reduced; at the same time, plants reach the end of their growing season and so transpiration also diminishes. In the spring, both processes become active again as temperatures rise.

We have already mentioned that it is possible to distinguish between surface- and ground-water regimes using this method. To do so, two or three holes are augered to different depths, a few metres apart. Thus, one hole may be drilled to a depth of 30 cm, another to 60 cm and the third to 100 cm. If the drainage impedance is due to ground water, the deepest hole will be the first to contain water, and the shallow ones will only become wet when the water table rises considerably. Conversely, if the impedance is due to surface water conditions, then the shallow holes will become water-logged (depending upon the depth of the impeding layer). In the example in *Figure 3.7* a compacted layer at 25–35 cm is causing surface-water impedance. A perched water table thus develops, and it is picked out by the shallowest hole; the deeper holes both penetrate the impermeable layer into the underlying freely drained soil, and therefore help to drain the soil.

Although the general principle of measuring the water table is very simple, one practical problem does occasionally occur. Where the water table is deep, it may not be possible to see the water in the hole. Accurate measurement of the depth is therefore difficult. The simple solution to the problem is to tape a length of absorbent cloth to the back of the rule or tape. When this is inserted into the water a wet mark will be left on the material and thus allow the water table to be defined. A more sophisticated probe can be constructed, however, so that when the tip of the probe touches the water an electrical circuit is completed and a bulb lights up. Details of how to make this are given in Appendix 1.

3.3 HYDROLOGICAL PROCESSES

3.3.1 Oxidation and reduction

We discussed many of the activities of water in physical weathering in Chapter 2. The role of water in leaching and chemical translocation we will consider when we come to deal with chemical processes in the soil. At this stage, however, it is useful to look at the processes of **oxidation** and **reduction**, for as we remarked earlier, these affect the colour of the soil and provide morphological evidence of drainage conditions.

When the soil becomes saturated, all the pore spaces in the soil are filled with water. As a result, the free entry of oxygen into the soil is limited, and since the diffusion of oxygen through water is very slow, inadequate amounts are available to meet the requirements of the soil organisms and plant roots. Oxygen is needed by living organisms for respiration; in simple terms, **electrons** (negatively charged particles) are produced during respiration, and combine with **oxygen ions**. If oxygen is unavailable, other ions in the soil must fulfil this role of accepting electrons produced during organic respiration. Iron and aluminium, in particular, accept electrons in the absence of oxygen. Since these are positively charged ions, whereas electrons have a negative charge, the result is that the net charge (valency) of the aluminium and iron is reduced; for example, in the case of iron

$$Fe^{3+} + e^- \longrightarrow Fe^{2+}$$

(Trivalent   (Electron)   (Divalent
iron)   iron)

This is an example of the process of reduction.

Since both iron and aluminium normally occur in the soil as oxides (that is, in combination with oxygen ions), the reduction of their valency in this way means that they lose an oxygen ion (O^{2-}). Clearly, three oxygen ions can combine with every pair of Fe^{3+} ions, but only two can combine with each pair of Fe^{2+}. Thus, during reduction, oxygen is often released from the iron or aluminium oxides:

$$2Fe_2O_3 \longrightarrow 4FeO \quad + \quad O_2$$

(Trivalent;   (Divalent;   (Oxygen;
ferric oxide)   ferrous oxide)   used by soil organisms)

In the ferrous state the iron oxide is characteristically dull grey in colour. Aluminium oxides, going through the same reaction, become greenish blue in colour.

By contrast, when the soil dries out, the larger pore spaces become free of water and air is able to enter. The electrons produced during organic respiration can then combine with the oxygen, and there is a tendency for the ferrous compounds to lose an electron. In this way, their valency increases, and they return to their ferric state. **This is the process of oxidation**. During this process the reactions outlined in the equations above proceed in the opposite direction. The reactions produce rust-coloured compounds of iron and aluminium, which give the soil a characteristic red or yellow colour.

In many soils, reduction and oxidation occur alternately as the soil is first wetted and then dried. An interesting distribution of mottles then develops. When the soil is saturated, the zone of maximum reduction occurs in root channels and larger pore spaces where the soil organisms are most active. Thus, these areas are picked out by intense gleying. Conversely, when the soil is dried, these same channels and pore spaces act as the main routes for oxygen movement, and thus they experience the most active oxidation. They are then defined by clear red and yellow mottles.

3.4 AGRICULTURAL IMPLICATIONS

3.4.1 Water and crop growth

Water is vital to plant growth for two main reasons. It helps keep plants moist, to maintain their **turgidity**; it also acts as a source of nutrient elements. If the available supply of water is less than their requirements, plant growth is reduced. They are then said to be suffering from a **soil moisture deficit**.

As long as the moisture content is above wilting point, plants are able to extract water from the soil. They do so in two main ways, either by 'attracting' water to the roots — a result of capillary forces — or by extending their root system in search of water. Movement of water to the roots is most important in soils with a large number of capillary pores; as roots extract the water from the pores the moisture film around the particles becomes thinner and thus a tension gradient is set up. Water then moves from the wetter soil towards the drier areas around the roots.

Movement by capillary processes is very limited. In most cases it only occurs over a distance of a few millimetres, and is too slow to provide plants with all the water they require. The extension of roots in search of water is also vital, therefore. It is at first astonishing to realise that plant roots may extend at a rate of several centimetres per day. However, anyone who has kept a particularly active house plant in a small pot, or tried to control a garden infested by thistles, may well believe that root growth can be prolific! Because of this rapid growth, plant roots are able to enter new areas of soil in search of water. Through the effect of capillary movement of water towards the roots they are then able to exploit an area considerably greater than that of the roots themselves.

If the plants are unable to obtain water from the soil, wilting eventually occurs. Before this, they show signs of **water stress**, with reduced rates of growth and a tendency to become limp. The reduced growth, however, is often due not so much to the lack of water as such, but to a depleted supply of nutrients. In the case of agricultural crops the result is reduced yields — a situation which is familiar to anyone who lives in areas liable to drought.

We have discussed the problems of lack of water, but it must also be remembered that plants, like humans, can have too much of a good thing. Excess moisture may be as detrimental as insufficient water. This is because when the soil becomes waterlogged, air is excluded and the supply of oxygen is limited. Since plant roots require abundant oxygen supplies, growth is seriously affected. Even a few days of saturation at critical stages in the growing season (especially during the seedling and flowering stages) may irreparably damage cereal crops. To obtain maximum yields it is therefore essential to remove excess water as quickly as possible. This introduces the principles of agricultural drainage.

3.4.2 Agricultural drainage

Purpose of drainage

Through the removal of excess water, soil conditions may be improved in a number of ways. Firstly, removal of excess water reduces the danger of damage to crops, by minimising the length of time in which the soil within the rooting zone may be saturated. In addition, it helps cultivation by increasing the rate at which the soil dries out after rainfall. Thus, the soil more rapidly achieves a state in which it is suitable for cultivation, at, or slightly below, field capacity. The soil structure is then sufficiently stable to withstand compaction or smearing by tillage implements. This is particularly important on clay soils which are inherently susceptible to structural damage when wet.

Drainage is also beneficial through its effect on soil temperature. A dry soil warms up more quickly than does a wet one because air is a poorer conductor of heat than is water. Consequently, during the spring, dry soils reach the critical temperature at which plant growth starts (*c*. 5°C) sooner than do wet soils. The growing season — the period when mean daily temperatures are above 5°C — is therefore longer on a dry soil, and plant growth is correspondingly prolonged (*Plate 6*).

As well as these direct effects upon plant growth, drainage has a number of indirect benefits. It improves the soil as an environment for micro-organisms and in this way encourages organic nutrient cycling. It also alters the chemical environment of the soil, and affects the weathering of soil minerals.

Methods of drainage

The method of drainage depends upon two main factors; the soil type and the requirements of the crop which it is intended to grow. As a general rule, it is not worthwhile putting in expensive systems of agricultural drainage if the returns, in the form of increased crop yields or quality, are insufficient to cover the costs. Consequently, relatively simple forms of drainage are used in association with low-value crops, such as rough pasture, while the systems become more sophisticated as the value of the crop increases.

Plate 6. Drainage and liming improve soil fertility and allow the development of more nutritious grasses; the improved pasture (arrowed) contrasts with the surrounding moorland vegetation (Photo D. Briggs).

Soil type is important in two ways. The physical properties of the soil — in particular its texture — control the range of possible drainage techniques, while the character of the drainage problem determines the type of improvement which is necessary. In general ground-water impedance is relatively easy to solve, for a system of sub-surface drains is able to carry away the excess soil water. So long as the water can then be removed from the area — by pumping if necessary — the drainage will be improved. Where poor drainage is a result of surface-water impedance, however, the problem is more difficult, for the impermeable layer will have a naturally slow hydraulic conductivity and thus the water will move very slowly through the soil into the drain system.

The implication of this is that drainage requires both a means of removing the excess water from the soil, and a means of moving it through the soil to the drainage system. In many cases, therefore, a dual system of drainage is required. We will briefly describe some of the main techniques.

(a) *Open ditches:* The simplest form of drainage is a series of open ditches or dykes. These are cut to a depth such that when the water drains into them, the resulting water table in the soil is below the rooting depth of the plants.

This system is only effective where the movement of water through the soil is rapid. It is therefore ideal in coarse-textured or well-structured soils in which impedance is due to ground-water conditions.

(b) *Mole drains:* As the name implies, mole drains consist of sub-surface channels in the soil, similar in form to mole tunnels. They are cut with a device shaped somewhat like a bullet which is dragged behind a tractor. The tunnels formed by the mole are only temporary — they last for up to 5–10 years — but they act as conduits for water below the soil surface. It is necessary to renew the moles as they collapse and become infilled.

Moling is generally carried out at a depth of 50–70 cm; if shallower they tend to be damaged by cultivation; if much deeper water takes a long time to reach them and drainage is ineffective. The opportunity for moling is strictly limited by the nature of the soil. It can only be carried out in fairly clayey soils, for otherwise the mole tunnels collapse too easily. It must also be carried out when the soil is close to or below field capacity, in order to avoid smearing of the soil. In many cases mole drains are used in association with a system of pipe or tile drainage.

(c) *Tile or pipe drainage:* In this system, a permanent network of subsurface drains is formed by cutting a trench and laying either clay tiles or plastic pipes. The ditch is then infilled, occasionally with a backfill of porous gravel or sand to help movement of water into the drains. This system of drainage is the most expensive commonly used, but new techniques are being devised which allow the drains to be installed in very narrow slits cut by the same machine. The technique is particularly suitable where the hydraulic conductivity is slow and a means of speeding up water movement through the soil is required. Commonly, a network of main and feeder drains is constructed, linked by a series of mole drains.

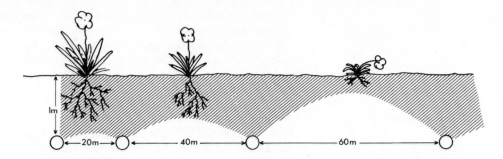

Figure 3.8 The effects of drain spacing on the soil water table

General principles

Whichever system of drainage is installed a critical factor is the depth at which it is installed. If it is too deep, the water level in the soil may be reduced too far, and the soil may tend to become droughty; plants will suffer from insufficient water. On the other hand, if the drainage system is too shallow, then damage to plant roots may still occur during wet seasons. In this context, it is important to realise that the water table has a slightly humped profile between the drains (*Figure 3.8*). As the spacing of the drains is increased, so the height in the middle of the 'hump' rises. Thus, the situation may occur in which the soil is adequately drained close to the drains, but suffers from water-logging between them.

A further point which must be noted is that the moisture conditions are liable to vary from year to year as the rainfall inputs change. It would be exorbitantly expensive to ensure that all the water was removed sufficiently quickly during the wettest imaginable period and, in any case, this would lead to lack of soil moisture during the drier seasons. Instead, we must consider the probability of certain conditions, and plan to improve the drainage to cope with rainfall levels which occur sufficiently often to make them a significant hazard. It is of course fortunate that for most crops one is only concerned with removing excess water during the spring to autumn period, when either the soil is being cultivated or a crop is being grown. During the winter the excess water is less important, and for the most part can be ignored.

Finally, it must be stressed that the aim of agricultural drainage is not to remove all the water from the soil. It is to control the drainage in order to create optimum moisture conditions for plant growth. Thus, drainage design is a skilled operation; a careful balance must always be maintained between retention and removal.

3.5 CASE STUDIES

3.5.1 Introduction

Because water is so important to plant growth, it is not surprising that many studies have been carried out comparing the hydrological properties of soils under different crops or different vegetation systems. At a very simple level there is the comparison of moisture contents in a bare fallow soil and a soil carrying a barley crop after a prolonged drought (*Figure 3.9*). This shows the importance of transpiration in removing water from the soil, particularly at depth. In the fallow soil practically no water was lost from below 45 cm.

In *Figure 3.10* the infiltration capacity of soils under a variety of land uses are compared. It is clear that factors such as increased trampling by livestock or compaction by tillage significantly reduce the infiltration capacity of the soil (*see Plate 2*).

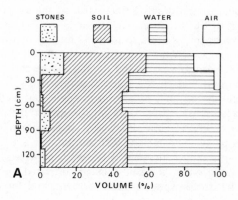

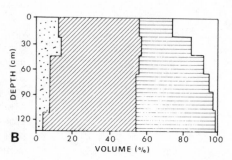

Figure 3.9 Soil moisture contents under a (A) fallow and (B) cropped surface (from Russell, 1973)

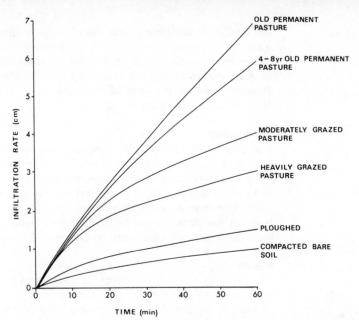

Figure 3.10 Land use and infiltration capacities (from Chow, 1964)

The figure shows infiltration rate (cm) on the vertical axis against time (min) on the horizontal axis, with curves labelled:
OLD PERMANENT PASTURE; 4 – 8yr OLD PERMANENT PASTURE; MODERATELY GRAZED PASTURE; HEAVILY GRAZED PASTURE; PLOUGHED; COMPACTED BARE SOIL.

3.5.2 Moisture regimes of clay soils (Thomasson and Robson, 1967)

The problem

Although it is frequently assumed that there is a close relationship between the soil moisture regime of a soil and its morphology – in particular its degree of gleying – there seem to be several instances where this relationship is only weakly expressed. In such situations it is difficult to predict to what position the water level will rise during the winter and fall during the summer; similarly, it is impossible to assess the likelihood of water-logging within the rooting zone during the growing season.

Soils developed from the red-brown clays and marls of the Permian and Triassic rocks provide a clear example of this problem. The bright colours of the parent materials, due apparently to a high proportion of ferric oxides in the clays, impart a dominantly red or reddish brown colour to the soils. These colours tend to mask the effects of weathering, and even in poorly drained sites the degree of mottling is subdued. Consequently to assess the drainage characteristics of these soils and to determine the agricultural significance of water-logging it is necessary to carry out detailed studies of the soil water regime. In this investigation, Thomasson and Robson measured the level of the water table in soils developed from the Keuper Marl in Nottinghamshire.

The method

Five sites were selected near Kingston-on-Stour. Site I and site II were in the same field, and the remaining sites (III–V) were within an area of 2.6 km². At each site three auger holes, of 10 cm diameter, were drilled to depths of 30 cm, 61 cm and 114 cm. The upper 30 cm of each hole was lined with a tile land drain, and the holes were covered with steel plates. At intervals of approximately 10 days the levels of the water in the holes were measured. At sites III, IV and V observations were continued for two years; at sites I and II measurements ceased after one year.

Results

The five sites showed a generally similar pattern of water-logging (*Figure 3.11*). The holes were dry for most of the summer, but the water table rose to within 100 cm of the soil surface throughout most of the winter. A slight discrepancy was seen between the holes drilled at different depths, the water level normally being highest in the 60 cm hole. In general, the water level also showed a close relationship to the pattern of rainfall during the preceding period, and water was present in the 30 cm holes, for example, only after prolonged or heavy rainfall.

There was one exception to this pattern. At site V, which was under woodland, the holes remained dry throughout the winter. This is presumably due in part to the greater transpiration losses of the trees compared to the grass or cereal crops at the other sites. However, it is also likely that the deeper roots of the trees have created a more permeable subsoil structure and thus improved drainage.

At sites III and IV the holes were occasionally pumped dry in order to measure the rate of recovery of the water table. At both sites recovery took over 2 weeks,

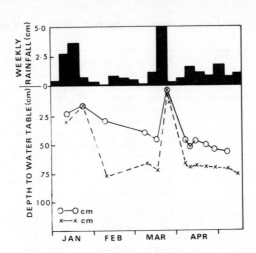

Figure 3.11 Water table levels in soils developed from Keuper Marl (from Thomasson and Robson, 1967)

in the 60 cm and 114 cm holes, showing the very low hydraulic conductivity. This reflects the heavy texture, small pore spaces and poorly developed structure of these soils.

During the late autumn of 1964, the field containing site IV was subsoiled at a depth of 76 cm. Following this treatment, the drainage of the soil improved noticeably. Whereas during the previous year site IV had been water-logged far longer than site III, after subsoiling the position was reversed. At site III the water level remained within 51 cm of the surface for 12–13 weeks; at site IV it reached this level for less than seven days.

Conclusions

Three interesting points come out of this study. In the first place it is clear that in the soils developed from the Keuper Marl drainage is sufficiently impeded to produce a permanent water table within the profile during the winter. Since the water level in the 61 cm hole was generally higher than that in the deeper hole, it

TABLE 3.2 TYPICAL PROFILE IN THE WORCESTER SERIES (from Thomasson and Robson, 1967)

0–20 cm	Dark brown (Munsell colour 7.5YR 4/2); clay loam; occasional stones; moderately developed medium subangular blocky structure; sharp boundary
20–41 cm	Reddish brown (5YR 4/4) and brown (7.5YR 4/4) faintly and finely mottled with yellowish red (5YR 4/6) and dark reddish grey (5YR 4/2); clay loam to clay; moderately developed medium angular blocky structure; narrow boundary
41–61 cm	Reddish brown (5YR 4/4); clay to silty clay; strongly developed coarse angular blocky structure with clay skins on ped faces; merging boundary
61–81 cm	Reddish brown (5YR 4/4); silty clay; calcareous; moderately developed coarse angular blocky structure; merging boundary
81 + cm	Reddish brown (5YR 4/4) with light olive grey (5Y 6/2) patches; silty clay loam; calcareous; strongly developed fine angular blocky structure

appears that the main impermeable horizon occurs at about 60 cm; the 114 cm hole apparently penetrated this layer into an underlying, more permeable subsoil. The drainage of the soil is therefore controlled by the rate at which water can pass through the poorly structured subsoil (*Table 3.2*).

The second feature of these results illustrates this point. After subsoiling, when the impermeable horizon was broken up by deep ploughing, the drainage improved markedly.

The third point of importance is the effect of the vegetation upon the moisture regime. Woodland clearly has a greater transpiration demand than grass or cereal crops, but it is rare for the effect to be so marked. The explanation is apparently that the impermeable layer is just below the active root layer of the agricultural crops, but well within reach of the tree roots. These were therefore able to penetrate the restricting horizon and accelerate the rate of drainage.

Comment

This study illustrates well the type of problem which measurements of soil water regime help to tackle. It is possible to use this method to identify impermeable horizons in the soil and thus to determine the most suitable corrective measures. It is also possible to assess the effects of artificial drainage by comparing water table levels in drained and undrained soils. Finally, it is clearly an appropriate technique for comparing the effects of different types of vegetation upon the soil moisture regime.

3.5.3 Effects of burning on the infiltration capacity of soils
(Scott, 1956)

The problem

Burning of the vegetation is used as a means of ecological and agricultural control in a variety of environments. Corn stubble is often burnt after harvesting to remove the unwanted plant debris and produce a clean soil surface. Heather moorlands are frequently burnt to encourage heather regeneration and provide a suitable habitat for grouse. Under conditions of shifting cultivation fire is used to prepare clearings for temporary settlement and cultivation. In many upland areas of woodland or brush burning is employed to clear areas for grazing.

In all these cases the fire may have a profound effect upon the soil. Almost certainly it affects the chemical status of the soil by removing some nutrients in the smoke, and by releasing others in sudden and large quantities from the vegetation. It also seems likely that burning influences the physical and hydrological properties of the soil. Several studies have suggested that the structure in particular changes when the soil is heated, and it has been argued that this may lead to a reduction in the infiltration capacity of the soil. In addition, ash produced by burning may block up the pores in the soil surface and further reduce its ability to absorb water. This investigation was aimed at measuring the effects of forest burning on the infiltration capacities of upland soils in California.

The method

Six areas were tested before and after burning in one year; during the following year five of the sites were revisited and a new site also tested. Infiltration capacities were measured with a simple cylinder infiltrometer like that described earlier in this chapter. Measurements were made over two time periods — 15 min and 1 h — and 10–24 replicate analyses were made at each site.

Results

The results of the investigations are given in **Table 3.3**. Two points are immediately clear. First, the infiltration capacity measured over a period of 15 min is in every case faster than that for a 1 h period. This reflects the tendency for infiltration to slow down as the soil becomes wetter; thus the 15 min reading is a measure of the instantaneous infiltration capacity, while the 1 h reading is more akin to the saturated capacity. Secondly, the burned plots generally have a higher infiltration capacity than the unburned. This is in contrast to the initial hypothesis.

Inspection of the results also shows that the repeat measurements at the five sites in the second year generally showed a slight increase in the infiltration capacity. This occurred in both the burnt and unburnt areas, and cannot be attributed to effects of regeneration of the vegetation or recovery of the soil structure. It seems more likely that it is a result of different soil moisture conditions at the time of analysis.

Conclusions

Clearly the most interesting aspect of this study is its implication that burning did not reduce infiltration capacities of the soil. Instead, it seemed to lead to a significant increase in the ability of the soil to absorb water. This is in contrast to the hypothesis previously established.

Part of the explanation is apparently that the ash produced by burning of the trees produced a porous surface layer, with an inherently high infiltration capacity. It also seems likely that the detrimental effects of burning tend to be cumulative. Repeated fires eventually lead to a loss of organic matter from the soil and the

TABLE 3.3 INFILTRATION CAPACITIES IN BURNED AND UNBURNED FOREST PLOTS
(all results in cm h^{-1}) (from Scott, 1956)

Site	Unburned		Burned	
	Year 1	Year 2	Year 1	Year 2
15 min				
1	4.7	13.4	5.9	10.1
2	37.4	44.8	45.9	49.2
3	12.3	36.5	34.9	44.5
4	13.6	16.9	35.9	21.1
5	18.6	16.5	61.1	38.9
6	19.2	–	23.5	–
7	–	49.4	–	53.6
60 min				
1	3.0	10.0	3.7	5.1
2	18.2	24.6	19.5	22.3
3	17.1	16.1	12.3	21.3
4	8.7	11.7	17.0	14.4
5	13.4	12.2	23.7	23.0
6	14.0	–	15.1	–

development of an unstable structure, at which point the infiltration capacity will be reduced. In the short term, however, the sudden inputs of plant debris from the vegetation may actually increase the organic content of the soil and improve the structural stability. In other words, to evaluate the full effects of fire it would be necessary to measure change in soil conditions after several years of repeated burning.

Comment

As we suggested at the beginning of this case study, fire is a common tool in man's management of the environment, and it obviously has wide-ranging effects upon the soil. Thus, measurements of infiltration capacities are not the only techniques which can be used to study its implications; it is also likely that analyses of chemical, physical and biological properties would reveal interesting and perhaps alarming results. In Britain, fire is used in two main situations: to remove stubble and to control heather. Both situations provide suitable conditions in which to analyse the effect of burning, either by comparing conditions before and after the fire, or, more generally, by analysing adjacent burnt and unburnt areas. It is not necessary, therefore, to set fire to the nearest patch of vegetation in order to find an appropriate study area!

One further lesson can be learned from this example. It may be noted that as many as 24 replicate measurements of infiltration capacity were made at each site. This reflects the inherent variability of infiltration capacities over quite small areas; to obtain anything like an overall impression of the infiltration properties of the soil, it is necessary to carry out a large number of replicate analyses. Single measurements, at one point, are of limited value.

CHAPTER 4 CHEMICAL PROPERTIES

4.1 INTRODUCTION

4.1.1 Soil nutrients

The nutrients in the soil can be broadly divided into two groups. One consists of those nutrients which are required by plants in large amounts, including hydrogen, carbon, oxygen, nitrogen, phosphorus, potassium, calcium and magnesium. These are generally known as **macro-** or major **nutrients**. The other group — the **micronutrients** — is required by plants in only small quantities and comprises a wide range of elements including sodium, iron, aluminium, silica, copper, molybdenum and zinc. This group therefore includes what are often referred to as trace elements.

The nutrients are important for a variety of reasons. **Carbon, oxygen** and **hydrogen**, for example, are the major constituents of plant tissue. **Nitrogen** is an important component of chlorophyll — the substance which enables plants to carry out photosynthesis (the conversion of sunlight to energy). Nitrogen is also found in protein, which is essential to cell growth and tissue renewal. **Phosphorus** occurs mainly in the protoplasm of the plants.

Plant growth depends upon the correct balance of these various nutrients. If the correct quantities are not supplied, then plant growth is impaired. This may occur if individual nutrients are present in either too small or too large amounts; thus toxicity, as well as deficiency, may limit plant growth.

4.1.2 Factors affecting soil nutrients

Weathering

The plant nutrients occur in three main forms in the soil — as ions bound up within the mineral particles, as ions adsorbed on to the surface of the colloids, and as ions in solution within the soil water (*Figure 4.1*).

During weathering, the soil minerals slowly disintegrate and decompose. In the process, individual ions are released from the minerals and are dissolved in the soil water. At the same time, weathering produces clay-sized particles (**colloids**) which have a small electrical surface charge. Ions released by weathering are attracted by these charges, giving a 'swarm' of adsorbed ions around each colloidal particle. Ions

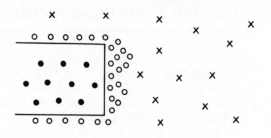

● Ions held in colloidal particles (unavailable)

o Ions adsorbed on colloidal surfaces (exchangeable)

x Ions in solution (available)

Figure 4.1 The distribution of nutrients in the soil

with a positive charge (**cations**), such as calcium, magnesium and potassium, are attracted to negative charges on the colloids; ions with a negative charge (**anions**), including phosphate, chlorides and sulphates, are attracted to positive sites on the colloids.

Of the three forms of nutrients, those in solution are the most readily available to plants and are therefore extracted from the soil. In addition, the dissolved nutrients may also be lost in drainage waters. As a result a gradual withdrawal of nutrients takes place which must be replenished by continued release of ions by weathering of the soil minerals. Clearly, therefore, the rate and type of weathering act as an important control on the nutrient content of the soil.

Parent materials and soil minerals

It is also apparent that the initial supply of nutrients is constrained by the character of the mineral particles and, hence, by the parent materials of the soil. In general, very siliceous, sandy rocks and soils provide a very limited quantity of nutrients, partly because they are weathered only slowly and partly because they contain a very limited range of nutrients. Conversely, clayey and calcareous parent materials produce a far wider range and greater quantity of nutrients.

Clays, in particular, have an important influence upon the nutrient status of the soil. Three main types of clay occur in the soil. The simplest form, including **kaolinite** and its related types, consists of sheets of oxygen atoms linked to sheets of silica and aluminium atoms. One sheet of silica is present for every sheet of aluminium and thus the clays are known as **1:1** clays.

A more complex structure is shown by **montmorillonite** and its associated types. This is formed of two sheets of silica to every sheet of aluminium atoms, each group of sheets being bound together by oxygen atoms. These are known as **2:1** clays.

The third group of clays are intermediate in form, a mixture of these two types occurring as inter-layered structures. These are often called interstratified clays, and they include **muscovite** and **illite**.

Of the three types of clay, **the 2:1 clays have the greatest number of surface charges** — a result of their more complex structure and chemistry. Thus they have the greatest ability to retain and adsorb nutrients on their surface. In contrast, the 1:1 clays have a very small surface charge, and therefore are able to adsorb only a limited number of nutrient ions. The interstratified clays are again intermediate in character (*Figure 4.2*).

Organic properties

Many of the nutrients removed from the soil by plants are eventually returned in the organic residues formed when the plants die. As we will see later, decomposition of these residues releases these nutrients, which are then further altered and recycled by soil organisms. Organic residues also break down to form colloidal particles which, like clay colloids, can hold ions by adsorption. Their ability to do so, however, is even better developed than in the 2:1 clays; consequently the content of organic matter in the soil greatly influences the retention and release of nutrients.

4.2 TECHNIQUES

4.2.1 pH

Introduction

One of the most abundant ions, both in solution and adsorbed on the colloidal surfaces, is hydrogen. This owes its abundance to the fact that it is one of the major constituents of water; a constant supply of hydrogen ions therefore enters the soil in rain-water. In addition hydrogen ions are released into the soil by plant roots.

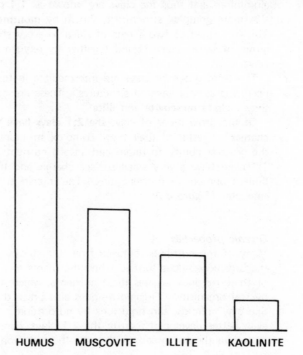

Figure 4.2 Relative quantities of nutrients adsorbed by different colloids

HUMUS   MUSCOVITE   ILLITE   KAOLINITE

Water is composed of molecules of two hydrogen ions linked to a single oxygen:

$$-O-$$
$$H^+ \qquad H^+$$

The individual ions have a tendency to *dissociate* — to break away — from the water molecules, producing two separate ions, one of hydrogen (H^+) and one of hydroxyl (OH^-):

$$H_2O \longrightarrow OH^- + H^+$$

The hydrogen ions released in this way can be adsorbed on to the colloids in the same way that other cations are. However, because of the plentiful supply of hydrogen ions in the soil water, they tend to be more numerous than cations such as calcium, magnesium and sodium, which are largely derived from the weathering of rocks.

The abundance of hydrogen ions in the soil water and attached to the soil colloids makes the soil acid (an acid is defined as a substance which dissociates to produce hydrogen ions). The more abundant are the hydrogen ions, the more acid is the soil. Thus if we measure the concentration of hydrogen ions in the soil we have a measure of its acidity. We make this measurement by reference to the pH scale.

pH is defined as the negative logarithm of the concentration of hydrogen ions in a solution, measured in moles per litre (a mole is the molar weight of a substance in grams):

$$pH = -\log_{10}[H^+]$$

This scale is used not only to measure the acidity of soil; it is also used to analyse chemical solutions. In the case of chemical solutions it has a range from 1.0 (very acid) to 14.0 (very alkaline). Neutrality is taken to be equivalent to pH 7.0. It must be stressed that because the pH is the **negative** logarithm of the hydrogen ion

TABLE 4.1 pH, HYDROGEN ION CONCENTRATION AND HYDROXYL ION CONCENTRATION

pH	H^+ concentration (moles l^{-1})	OH^- concentration (moles l^{-1})
3.0	0.001	0.00000000001
4.0	0.0001	0.0000000001
5.0	0.00001	0.000000001
6.0	0.000001	0.00000001
7.0	0.0000001	0.0000001
8.0	0.00000001	0.000001
9.0	0.000000001	0.00001

concentration, a high concentration of hydrogen ions in the solution is represented by a low pH value. Conversely, when the H^+ concentration is low, the pH value is high. Thus, **acids have a low pH**, while **alkalis have a high pH**.

It should also be noted that an inverse relationship exists between the concentration of hydrogen ions and hydroxyl ions in the solution. At a pH of 7.0 there are equal concentrations of both H^+ and OH^-, and hence the solution is considered to be neutral. This is the situation which exists in pure water. As the pH falls, however, the concentration of hydrogen ions increases while that of the hydroxyl ions decreases. Because the pH scale is calculated to the base of ten, each 1.0 fall in the pH is a reflection of a tenfold increase in the concentration of hydrogen ions in the solution, and a tenfold decrease in the concentration of hydroxyl ions. Thus, at pH 6.0 the concentration of hydrogen ions is 100 times that of hydroxyls; at pH 5.0 the ratio is 10 000:1 (*see Table 4.1*).

It is clear from what has been said that pH is a measure of the concentration of hydrogen ions in solution; it does not measure the total quantity held on the colloidal surfaces. The question may therefore arise: how are we actually measuring the acidity of the soil? The answer is that there is an intimate relationship between the ionic concentration in the solid and liquid phases of the soil. As we will see later, if an imbalance exists, ions are released from the colloidal surfaces and replaced by ions dissolved in the soil water. This exchange continues until equilibrium is established. Thus, if we measure the concentration of hydrogen ions in the soil solution we are, in fact, obtaining a measure of the acidity of the total soil.

We may mention finally that although the theoretical pH range is from 1.0 to 14.0, in soils the actual range of values is generally from about 3.0 to 10.0. Examples do occur when the soil pH lies outside these limits, but these are the result of special circumstances.

Measurement

ELECTROMETRIC METHOD:

Equipment: pH meter
2 100 ml beakers

	Stirring rod
	50 ml measuring cylinder
	Balance
Reagents:	Buffer powders
	Distilled water or 0.01M $CaCl_2$
Sample:	*c*. 20 g fresh soil
Procedure:	Switch on the pH meter 30 min before use.

Weigh out 20 g fresh soil into a 100 ml beaker; add 50 ml distilled water or 0.01M $CaCl_2$ from the measuring cylinder.

Stir thoroughly and leave to stand, stirring occasionally, for 30 min.

Make up a buffer solution as directed on the packet (normally one tablet or powder in 100 ml distilled water).

Insert the electrodes of the pH meter into the buffer solution and turn the pH control to the 'read' position; adjust the reading of the instrument to the pH of the buffer solution.

Remove the electrodes, rinse in distilled water and insert in the soil suspension; read the pH to the nearest 0.1 pH units.

N.B. Between all readings return the pH electrodes to distilled water and switch the pH control to 'off'. Some machines also have a temperature control which must be adjusted to the temperature of the solutions being measured.

COLORIMETRIC METHOD

Equipment:	Test tube and bung
	Small funnel
	Colour chart

Reagents: Barium sulphate
             Indicator liquid
             Distilled water

Sample: 5–10 g fresh soil

Procedure: Pour fresh soil into the test tube to a depth of

             10 mm for clayey soils
             25 mm for loamy soils
             40 mm for sandy soils

             Add barium sulphate to a total depth of 50 mm.

             Add 20 mm distilled water.

             Add 10 mm indicator liquid.

             Insert the bung and shake thoroughly; stand the tube upright until a clear supernatant develops.

             Compare the colour of the supernatant with the pH colour chart (or *Table 4.2*); record the pH to the nearest 0.5 units.

TABLE 4.2 COLOUR OF SUPER-NATANT IN COLORIMETRIC TEST FOR pH

pH	Colour
1–2	Red
3–4	Red/orange
5	Orange
6	Yellow
7	Yellow/green
8	Green
9	Blue/green
10	Blue

Various methods of pH determination are available. The most accurate is the electrometric method, which involves measuring the electrical potential developed across a membrane between an alkaline reference solution and the sample solution. The alkali standard is contained in the reference electrode of the pH meter, and this is inserted in a suspension of soil and either distilled water or dilute calcium chloride (1.111 g dry $CaCl_2$ dissolved to give 1 l). Although the sample is normally made up in distilled water, there are several advantages of using a solution of calcium chloride. In particular, this helps stabilise the pH reading in calcareous or saline soils. It tends to produce a pH value about 0.5 units lower than measurements made in distilled water.

The reading is also affected by the soil:solution ratio used in preparing the sample. Clearly, as the volume of solution is increased, the sample becomes more and more artificial — less like the soil in its field state. The pH value obtained may therefore not be an accurate representation of the pH of the soil in the field. On the other hand, if very little solution is used, it may be difficult to obtain a good contact between the electrode and the soil solution. In addition, it is easy to damage the electrode. For this reason, although various concentrations may be recommended, probably the most widely used is a ratio of 1:2.5 (i.e. 20 g soil to 50 ml solution).

The high degree of precision obtained by using the electrometric method is obviously important in trying to detect subtle differences in soil pH. However, this level of precision may be misleading, since variations of considerably more than 0.1 pH units may occur in response to changes in soil moisture content or temperature. Thus, the pH of the soil may fluctuate both naturally over time, or during transport and storage. For this reason, it is generally more meaningful to measure the pH of the soil as soon after sampling as possible. Field pH meters are in fact available, which run off batteries, and allow the pH to be determined on site. These, however, are less precise than the laboratory instruments.

An alternative approach to pH measurement is the colorimetric method. This is based upon the fact that certain organic dyes are sensitive to changes in pH; as the pH changes the configuration of the molecules alters and so its colour changes. By combining a number of different dyes an indicator solution can be made which is sensitive to small changes over a wide range of pH. These are mixed with a suspension of soil and water, and barium sulphate is added to help the soil settle from suspension. The suspension is thoroughly mixed and allowed to stand until a clear supernatant develops above the soil. The colour of this gives the pH of the soil.

Complete kits are available, containing all the materials necessary for the colorimetric determination of pH. Alternatively, an indicator solution may be made up from the following reagents (Smith and Atkinson, 1975):

0.06 g	methyl orange
40.00 ml	methyl red
0.08 g	bromthymol blue
0.10 g	thymol blue

0.02 g phenolphthalein
100.00 ml ethanol
0.1M NaOH added dropwise until a yellow colour develops

This gives the colour reactions listed in *Table 4.2.*

The colorimetric test for pH is certainly accurate to 0.5 pH units and, by inter-
polating between colours it may be possible to determine pH to within 0.25 units.
Even so, this method clearly lacks the precision obtained by electrometric measurement. Nevertheless, it has the advantage that the soil may be analysed immediately,
in the field, and changes in pH caused by transport and storage are avoided. The
major limitation to the method is that it is not suitable for highly organic soils,
for the organic matter absorbs the dye and thus interferes with the development of
the indicator colour.

An even more simple method of pH measurement is available, based upon the
colorimetric technique. It is possible to obtain pH papers which, like the colorimetric
indicator solutions, are sensitive to small changes in pH. Thus a suspension of soil
and water is made up (*c.* 10 g soil to 25 ml water) and a pH paper is dipped into
the suspension. The pH paper will change colour according to the pH; from red in
very acid soils, to yellow in neutral soils and blue-green in alkaline soils. In this
case, however, it is only possible to determine pH to the nearest 1.0 units. Thus,
although the method serves as a useful spot test, it only picks out fairly large
changes in pH.

4.2.2 Calcium carbonate content

Introduction

Calcium carbonate occurs in the soil in a number of forms. In soils derived from
calcareous parent materials, such as chalk and limestone, it represents the partially
weathered residues of the source rock. In other situations, it is deposited in the soil
by percolating calcareous waters; $CaCO_3$ precipitates from solution owing to changes
in the pH and aeration of the soil or to evaporation of the soil water during dry
periods. It may also occur in the form of shell fragments derived from the tests of

snails and other organisms. Finally calcium carbonate is often applied to soil in the form of lime.

Small amounts of calcium carbonate are therefore present in a wide variety of soils. Its importance, however, is increased by the fact that it is very soluble, particularly in acidic water. Consequently, calcium carbonate is rapidly removed from the soil during weathering, and the content remaining in the soil may provide an index of the extent of weathering. Calcium is also an essential plant nutrient, and measurement of the $CaCO_3$ content gives an indication of its availability to plants. Thirdly, calcium has an important effect upon the soil pH. Calcium carbonate dissociates to produce calcium and carbonate ions:

$$CaCO_3 \longrightarrow Ca^{2+} + CO_3^{2-}$$

In the presence of free hydrogen ions, the carbonate ions combine to form bicarbonate, which dissociates to produce water and carbon dioxide

$$CO_3^{2-} + 2H^+ \longrightarrow H_2CO_3 \longrightarrow H_2O + CO_2$$

In this way the concentration of free hydrogen ions in the soil solution is reduced, and the pH of the soil raised.

Measurement

Equipment: Small crucible
             100 ml beaker
             Balance

Reagents: 2*M* hydrochloric acid: add 183 ml HCl to distilled water and dilute to 1000 ml

Sample: 5–10 g dry soil

Procedure: Weigh 5–10 g soil into the small crucible, record the weight as W_1.

Pour 20 ml 2*M* HCl into the beaker.

Balance the crucible on the beaker and weigh; record the weight as W_2.

Tip the soil into the acid; swirl to mix and leave until the reaction ceases; reweigh and record the weight as W_3.

Calculation: The $CaCO_3$ content of the soil is given by the equation:

$$CaCO_3 = (W_2 - W_3) \times \frac{227.2}{W_1}$$

Because calcium carbonate is readily soluble in acid, the quantity of $CaCO_3$ in the soil can be assessed by noting the effect of adding dilute acid (10% HCl) to a small soil sample. From the energy of the reaction, it is possible to estimate the $CaCO_3$ content (*Table 4.3*).

A more precise approach is provided by the gravimetric method of determination. In this case, the $CaCO_3$ is dissolved in 2*M* hydrochloric acid, and the weight loss during the reaction determined.

When hydrochloric acid is added to the soil, the calcium carbonate dissolves to produce calcium chloride, water and carbon dioxide.

$$2HCl + CaCO_3 \longrightarrow CaCl_2 + H_2O + CO_2$$

The carbon dioxide is given off as a gas, and is lost to the atmosphere. Consequently by determining the loss in weight resulting from the loss of carbon dioxide, it is possible to assess the original quantity of $CaCO_3$. Since each gram of $CaCO_3$ releases 0.44 g CO_2, it follows that every gram loss in weight during the reaction represents 2.272 g $CaCO_3$.

4.2.3 Lime requirement

Introduction
In general, nutrients are most readily available to plants when the soil pH is in the slightly acid range, that is, from about 6.5 to 7.0. Consequently, most agricultural

TABLE 4.3 ANALYSIS OF CaCO₃ CONTENT BY ACID REACTION (from Clarke, 1971)

CaCO₃ content	Reaction
<0.1%	None
0.1–0.5%	Faint spitting audible
0.5–1.0%	Spitting audible
1.0–2.0%	Clearly audible; slight reaction visible
2.0–5.0%	Easily audible and visible
5.0–10.0%	Vigorous effervescence

crops grow best when the soil has a pH of about 6.5. In Britain, however, many soils have a natural pH somewhat below this value — they are rather more acid than the optimum for plant growth. As a result, farmers frequently need to raise the pH of the soil in order to increase plant growth and obtain higher yields.

As we have seen, calcium carbonate tends to raise soil pH by removing excess hydrogen ions from the soil solution. For this reason, one of the most useful methods of increasing soil pH is to add limestone or chalk to the soil. Other substances may be used with similar effect, however. **Calcium oxide** (CaO) or **burned lime** as it is often known, may also be employed; so too may **calcium hydroxide** ($Ca(OH)_2$) which is often referred to as **slaked lime**. In all cases, the overall reaction is essentially the same. Calcium ions are released by dissociation, and hydrogen ions — which would otherwise render the soil acid — are combined with carbon and oxygen compounds to form bicarbonate. This then dissociates to produce water and carbon dioxide.

The effect of adding liming materials to the soil is not always the same, however. The change in pH varies according to the nature of the soil. For this reason, it is not possible to predict the effect of liming the soil without first carrying out experiments. To determine the **lime requirement** of the soil — that is, the amount of lime needed to raise the soil pH to 6.5 — it is therefore necessary to determine the change in soil pH resulting from known additions of lime.

Measurement

Equipment: 5 100 ml beakers
Balance
pH meter or colorimetric pH kit

Reagents: Calcium hydroxide ($Ca(OH)_2$)
Chloroform
Distilled water

Sample: c. 100 g fresh soil

Procedure: Weigh out 5 20 g samples of fresh soil and place in separate 100 ml beakers.

Add 50 ml distilled water to each sample.

Add 0, 0.01, 0.02, 0.05 and 0.10 g $Ca(OH)_2$ to the beakers.

To each sample add 2 drops chloroform.

Stir thoroughly; cover and leave for 4 days.

Stir; measure the pH of each sample.

Calculation: Draw up a curve showing the weight of calcium hydroxide added against the pH of the soil.

From the curve, determine the amount of calcium hydroxide necessary to raise the pH to 6.5; record as W_h.

Calculate the amount which would be needed to raise the pH of 1 hectare, to a depth of 20 cm, to a pH of 6.5;

$$W_1 = W_h \times 125$$

Calculate the quantity of liming material needed:

$$CaO = W_1 \times 0.76$$

$$CaCO_3 = W_1 \times 1.37$$

$$Ca(OH)_2 = W_1$$

Although calcium carbonate is probably the most common material used to counteract soil acidity, determination of the lime requirement under laboratory conditions is more easily carried out using calcium hydroxide. This is more soluble than $CaCO_3$ and therefore rapidly reaches equilibrium with the soil.

The method of analysis is essentially simple. A series of samples are treated with varying amounts of calcium hydroxide, and a few drops of chloroform are added to prevent microbial activity. The samples are then covered and left for 4 days, after which the pH is determined. A graph is drawn to show the relationship between final pH and amount of calcium hydroxide used, and from this the lime requirement of our sample can be determined (i.e. the quantity of calcium hydroxide needed to raise the pH to 6.5). Normally the curve shows an approximately parabolic form.

It is then necessary to convert this result into tonnes per hectare. This is achieved by multiplying the value obtained from our 20 g sample by 125. If necessary this can be further converted to give an estimate of the amounts of limestone ($CaCO_3$) which would be needed. To do so, we multiply by a constant of 1.37; this is the ratio between the amount of Ca in each tonne of calcium hydroxide and the amount in each tonne of calcium carbonate.

Probably the most serious difficulty in applying this method is that it involves making up a suspension of soil and water, to which the liming material is added. In reality, of course, the lime is added in a dry state to a more-or-less dry soil. Thus, it comes into solution very slowly, and may take many years to completely dissolve. The effect upon the pH therefore tends to be much less marked under field conditions than may be implied by laboratory analysis, and there is a tendency using this method to underestimate the quantity of lime needed. Nevertheless, for comparative purposes, it does have general value.

4.2.4 Nitrate content

Introduction

Of all the plant nutrients in the soil, possibly the most important in terms of plant growth is nitrogen; frequently its supply governs the yield of crops that have sufficient water. Not surprisingly, therefore, it is one of the most widely used fertilisers.

The nature of the nitrogen supply will be discussed in detail in the next chapter, for it is closely tied up with the activity of soil organisms. It is important here to summarise a few major points, however. Nitrogen is obtained primarily from the atmosphere and is fixed in the soil by micro-organisms and plants. When these die and decay ammonium ions (NH_4^+) are produced. Through the activity of soil micro-organisms this is rapidly oxidised, first to nitrite (NO_2^-) and then to nitrate (NO_3^-). In this form it is readily available to plants. To measure the amount of nitrogen available to plants, therefore, we determine the quantity which is present in the form of nitrate.

Nitrate is easily removed from the soil in drainage waters; as a result the amount of nitrate in the soil depends upon the balance between the rate of supply from plant decomposition and the rate of removal, either by percolating waters or by plants. In many well-drained, acid soils the nitrate content is extremely low and plant growth, and vegetation composition, are consequently affected.

Measurement

Equipment: Small funnel and filter paper
Teaspoon
10 ml test tube
Pasteur pipette
5 ml test tube

Reagents: Brucine solution: dissolve 0.5 g brucine in 100 ml chloroform containing 5 ml HCl. N.B. Brucine is poisonous.

Morgan's solution: dissolve 100 g sodium acetate ($NaC_2H_3O_2.3H_2O$) in about 500 ml distilled water, containing 30 ml conc. acetic acid; dilute with distilled water to 1000 ml.

Conc. sulphuric acid (H_2SO_4).

Sample:	5–10 g fresh soil
Procedure:	Make the filter paper into a cone and insert it in the funnel.

Measure out approximately 5 g (one level teaspoonful) of fresh soil into the filter paper.

Using the 10 ml test tube measure out 10 ml Morgan's solution and pour it on to the soil.

Allow the first few drops of filtrate to go to waste, then insert the 10 ml test tube below the funnel to collect the filtrate.

Using the pipette, put 3 drops of the filtrate into the 5 ml test tube.

Add 2 drops of brucine and 7 drops of conc. H_2SO_4; mix by swirling.

After one minute note the colour which develops.

The precise quantitative determination of nitrate-nitrogen in the soil is extremely complex, and much research is still directed to perfecting efficient techniques of measurement. For rapid, semi-quantitative assessments, however, a simple colorimetric method has been devised, which uses brucine as an indicator solution. This can be carried out in the field.

The nitrate is first extracted from the soil by allowing a volume of Morgan's solution (a solution of acetic acid and sodium acetate) to percolate through a small sample. The filtrate is collected and a few drops placed in a small test tube. Two drops of brucine and seven drops of concentrated sulphuric acid are then added and the solution agitated to mix it. After a few seconds a colour develops, dependent upon the nitrate content (*Table 4.4*).

TABLE 4.4 SOLUTION COLOURS IN THE COLORIMETRIC TEST FOR NITROGEN

Colour	Nitrate–nitrogen (kg ha^{-1})	Agricultural significance
Pink	20	Deficient
Pale yellow	50	Slightly deficient
Moderate yellow	75	Abundant
Straw yellow	100	Very abundant

This technique is ideally suited to rapid field work. In this case a series of standard solutions are prepared by dissolving potassium nitrate (KNO_3) in Morgan's solution (0.0361 g dry KNO_3 diluted in 500 ml Morgan's solution gives 50 ppm N). These are treated in the same way as the soil extracts and the colours which develop are compared. In this way it is possible to determine the nitrate content of the soil to the nearest 5–10 ppm.

It must be stressed that brucine is poisonous, and sulphuric acid may cause severe burns. The reaction when adding the sulphuric acid may be violent, and care must be taken when handling the reagents.

4.2.5 Phosphate content

Introduction

Phosphorus occurs in a number of different forms in the soil. In most soils the main source is from the weathering of minerals such as calcium phosphate (apatite) and iron or aluminium phosphates. Additional quantities are also supplied from organic material, and in very peaty or sandy soils this may be the major source. Neither the organic phosphorus, nor the inorganic, is readily available to plants, however, and therefore, despite its overall abundance in the soil, phosphorus is often deficient as a plant nutrient. For this reason, phosphatic fertilisers are widely used to maintain the supply to agricultural crops.

Plants take up most of their phosphorus in the form of phosphate (PO_4^{2-}). The exact nature of the phosphate varies according to soil conditions — in particular, the acidity — but we can obtain a measure of the potential supply of phosphorus to plants by measuring the quantity of phosphorus which is present in the form of PO_4.

Measurement

Equipment: Small funnel and filter paper
10 ml test tube
5 ml test tube
Teaspoon
Pasteur pipette

Reagents: Morgan's solution: make up as for the nitrate test.

Molybdate reagent: dissolve 15 g ammonium molybdate $((NH_4)_6Mo_7O_{24}.4H_2O)$ in about 300 ml distilled water; slowly add 500 ml conc. hydrochloric acid; cool and dilute to 1000 ml; store for up to 3 months in dark glass.

Stannous oxalate: dissolve 0.5 g stannous oxalate (or stannous chloride) in 100 ml dilute (10%) hydrochloric acid; warm to dissolve; filter into a small glass container and seal well; store for no more than 1 week.

Sample: 5–10 g fresh soil

Procedure: Fold the filter paper into a cone and place in the funnel.

Measure out 5 g (1 level teaspoonful) fresh soil and pour into the filter paper.

Pour 10 ml Morgan's solution from the 10 ml test tube on to the sample.

Allow the first few drops of filtrate to go to waste, then insert the 10 ml test tube below the funnel and collect the filtrate.

Using the pasteur pipette place 10 drops of filtrate into the 5 ml test tube.

Add 2 drops of ammonium molybdate; shake to mix; add 1 drop of stannous oxalate.

Shake gently to mix, then allow to stand for 1 min; note the colour which develops.

As is the case with nitrate, the determination of phosphate levels in the soil by accurate and quantitative methods is difficult, but simple field tests are available. Basically the technique is similar to that used for measuring nitrate levels; the soil is treated with Morgan's solution to extract the phosphate, and then the quantity in the extractant solution is determined by a colorimetric test. Ammonium molybdate and stannous oxalate are used to develop a blue colour in the soil extract, the intensity of which reflects the concentration of phosphate.

Again, the test may be made more quantitative by preparing standards of known phosphate content and treating them in the same way. For this purpose potassium dihydrogen phosphate (KH_2PO_4) is used; 0.43 g dry KH_2PO_4 dissolved in 1 l of Morgan's solution gives a standard of 100 ppm. This can be further diluted to give standards of 50, 25, 10, 5 and 1 ppm. The colour developed from these standards can then be used to determine the phosphate phosphorus content of the soil extract by visual comparison. It may also be noted that simple comparators are available from many manufacturers which allow a more precise comparison of solutions prepared in this way.

4.3 CHEMICAL PROCESSES

4.3.1 Solution

One of the basic principles of chemistry is that **all substances tend towards a condition of equilibrium with their surrounding environment**. It is in an attempt to achieve this equilibrium that soil particles undergo weathering; weathering can therefore be seen as an '**equilibrium reaction**' — a means of attaining equilibrium.

One of the main equilibrium reactions in the soil is **solution**. This involves the movement of materials from the solid to the liquid phase of the soil — from the minerals into the soil water. The fundamental stimulus for this movement is provided by a chemical disequilibrium between the composition of the solid and that of the liquid. Thus, minute particles will move from the solid into the solution, until equilibrium is achieved.

In reality, numerous factors affect this process. One of the most important considerations is the acidity of the water. In general, **minerals are more soluble in acid conditions**, and thus they are dissolved more rapidly. A second important consideration is the amount of carbon dioxide in the water. Carbon dioxide is dissolved from the atmosphere by rain-water, and is also released into the soil water by respiration of soil organisms and plant roots. This increases the ability of the water to dissolve many substances, in particular calcareous minerals such as calcite.

The rate of solution is also affected by the ease of removal of the dissolved substances. If the water in the soil is static, the films of water around the minerals will rapidly become saturated with material dissolved from the mineral surface. Further solution will only occur as this material diffuses away through the surrounding water. On the other hand, if the water is moving, such that fresh water is constantly being introduced to the mineral, solution will continue since a constant state of disequilibrium will be maintained. It is for this reason that solution of bedrocks is most active along cracks such as joints and bedding planes. It is also for this reason that solution is most intense in porous, well-drained soils.

4.3.2 Dissociation

Solution is often related to a second chemical process, namely **dissociation**. This refers to the tendency for compounds to split up into their individual parts. Calcium carbonate, for example, splits into its constituent calcium and carbonate ions:

$$CaCO_3 \longrightarrow Ca^{2+} + CO_3{}^{2-}$$

Dissociation may occur as part of the process of solution. Solids dissociate as they dissolve, so that they occur in the soil water in the form of individual ions. Dissociation may also occur independently, however, within the soil water. Thus, materials such as calcium carbonate or sodium chloride may dissociate while in solution. We have already noted this process with respect to the water molecules themselves; when discussing pH we saw that the H_2O dissociated into individual hydrogen and hydroxyl ions. It was in this way that hydrogen ions were released to make the soil acid.

4.3.3 Cation exchange

We have already noted (Section 4.1.2) that colloidal particles in the soil have the ability to attract ions on to their surfaces. This ability arises from the fact that during weathering, clay minerals develop a net surface charge. This is due to two processes. In the first place it occurs because of **isomorphous replacement**: ions within the structure of the clays are replaced by other ions of similar size and shape but different valency. In general, this process involves the replacement of ions of high valency by others of lower valency. Thus silica (Si^{4+}) may be replaced by aluminium (Al^{3+}); aluminium in turn may be replaced by magnesium (Mg^{2+}). The overall effect is to upset the electrical neutrality of the mineral and produce a net negative charge.

The second process is **breakage**. During weathering the clay minerals tend to fracture along weak structural surfaces, and this leaves 'loose ends' — unsatisfied charges — at the edge of the mineral. In this case both positive and negative charges may be produced.

The negative charges produced in these ways attract positively charged ions. Thus **monovalent** ions of potassium (K^+) or sodium (Na^+) may become attached to one of the unsatisfied charges, thereby neutralising it. Alternatively **divalent** calcium (Ca^{2+}) or magnesium (Mg^{2+}) may become attached to two negative charges, neutralising them both. Cations attached in this manner to the surfaces of the clay particles are said to be **adsorbed**. The charges to which they are held are known as **exchange sites**; under appropriate conditions, the adsorbed cations can be released and replaced by other ions.

The impetus for this exchange arises largely from a chemical imbalance between the surface of the colloidal particle and the surrounding soil solution. There will always be a tendency for the ionic concentration in these two phases to adjust towards equilibrium. Thus if, say, hydrogen ions are more abundant in the soil solution than on the colloidal surfaces, they will tend to move from the solution on to the clays, in the process displacing other ions held on the clay surfaces. This process will continue until equilibrium is restored.

Several other factors affect **cation exchange**. One of the most important is the nature of the colloidal material. As we saw earlier there are three main types of clay. The 1:1 clays have a very small surface charge, and are thus able to hold and exchange only small quantities of ions; they are said to have a low cation exchange capacity. The interstratified clays and, even more so, the 2:1 clays have much larger surface charges and therefore greater exchange capacities.

It has also been mentioned that organic residues have a very strong surface charge, resulting from the decomposition of carbon compounds. These therefore have exceptionally high exchange capacities, and where organic matter is present the soil has a very marked ability to adsorb and release ions.

Thus far we have assumed that ion exchange only involves cations. In fact **anion exchange** may also occur, but because the presence of positively charged exchange sites on the colloidal particles is relatively rare, this process is far less important.

It has also been implied that ion exchange only takes place between the soil colloids and the soil water. This is not entirely true, however. Exchange can occur between the soil water and the roots of plants, or directly between the roots and colloids. The plant roots tend to release hydrogen ions during these exchanges, and thus they act as a major source of soil acidity.

4.3.4 Leaching

The ions released into the soil water by the processes of solution, dissociation and cation exchange may follow a variety of paths through the soil. Some are taken up by plants; others are re-adsorbed on the colloidal particles. Yet others are washed downwards through the soil by percolating waters, a process known as **leaching**.

Leaching is most intense under conditions of high rainfall, free drainage and low

Plate 7. A podsol soil, Roxburghshire, Scotland. Note the peaty surface horizon, the underlying pale leached layer and the darker ironpan in the middle of the profile (Photo D. Briggs).

pH. The combination of an abundant and constant supply of acid waters and high solubility of the soil materials allows a wide range of nutrients to be leached. Under these conditions, many of the more soluble constituents are, in fact, removed completely from the soil profile, and as a result the soil becomes deficient in nutrients and very acidic. In less intense conditions, however, many of the solutes are precipitated in the subsoil, producing zones of enrichment (illuvial horizons). Thus, leaching of iron in podsolic soils (*Plate 7*) leads to the formation of ironpans in the subsoil; leaching of calcium in calcareous soils may produce zones of secondary calcite in the lower horizons. At the same time, of course, the upper horizons of the soil, which experience the greatest loss of soluble materials, are depleted of their nutrients and form eluvial layers.

It follows from this that the character of the soil profile, and in particular the nature of the soil horizons, closely reflect the chemical processes operating in the soil.

4.4 AGRICULTURAL IMPLICATIONS

4.4.1 Chemical aspects of soil fertility

We have already seen that plant growth depends upon the supply of a range of plant nutrients, many of which are derived from the soil. Clearly, therefore, the ability of the soil to supply these nutrients controls its fertility.

Numerous factors influence the availability of plant nutrients. In the case of those nutrients, such as nitrogen, supplied by biological processes of fixation and mineralisation, the biological properties of the soil are particularly important. In the case of minerals such as phosphorus and potassium, a more crucial factor is the rate of release of ions from their source minerals. This is a reflection of the rate at which a particular nutrient is released from **insoluble** 'reserve' forms, which are unavailable to plants, to **soluble** forms which are available to plants.

A pool of readily available nutrients occurs in the soil solution and these supply the plants directly. As these are withdrawn from the soil solution, more nutrients are released from the reserves held by adsorption on the colloidal particles. In time, however, this reserve supply will be depleted, and it will only be slowly replenished by the action of weathering. Thus, most nutrients occur in three forms:

1. **Available nutrients** – dissolved in the soil water.
2. **Exchangeable nutrients** – held on the surfaces of colloidal particles.
3. **Labile nutrients** – held within the soil minerals and released by weathering.

In the short term the rate of release of nutrients from the exchangeable reserve into the soil solution is sufficiently rapid to meet the needs of plants. The rate of replenishment of the exchangeable reserves by minerals from the labile pool is slow, however. Consequently, over time the exchangeable reserves tend to be depleted, and the soil becomes 'exhausted'.

The supply of plant nutrients and thus the fertility of the soil, is also affected by pH. **The solubility of most nutrients varies in response to pH**. Most become more soluble as pH falls, and thus they are released more rapidly in acid conditions. As

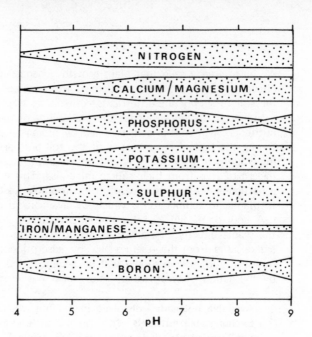

Figure 4.3 The relationship between pH and nutrient availability

acidity increases, however, the losses of these nutrients by leaching increase also, and thus their availability to plants may actually decrease. In other cases, the quantity of some nutrients may rise so greatly under acid conditions that they become toxic to plants.

The general relationship between the availability of nutrients and pH of the soil is given in *Figure 4.3*. It is clear from this that most nutrients are at maximum availability in slightly acid soils. In addition, we may note that soil acidity also affects the activity of soil organisms (*Figure 5.3*) — a factor which indirectly influences soil fertility. For these reasons, one of the main aims of soil management is to control the pH, to produce a slightly acid to neutral soil.

4.4.2 Management of chemical fertility

Liming

In general, British soils are somewhat more acid than the optimum for plant growth. Consequently, attempts to control pH commonly involve increasing the soil pH. This is normally done by adding an alkaline substance to the soil, lime being the most widely used material.

Adjusting the pH is not simply a question of adding a known amount of lime to a known volume of soil, however, for soils differ in their hydrogen buffer capacity — that is, their tendency to respond to changes in hydrogen concentration. Under some conditions, hydrogen ions may be released from the 'labile pool' associated with the colloidal particles and replenish those lost by exchange. As a result, the soil remains acid despite the addition of lime. This reserve acidity, as it is called, means that the pH of the soil will not always respond immediately, or in proportion, to the application of a liming material. For this reason, it is often necessary to carry out tests to determine the reaction of the soil to lime and, thereby, determine the quantity needed to raise the pH to 6.5.

Fertilisers

Although lime is probably the main material used in controlling the chemistry of the soil, many other chemicals may be used to adjust directly the supply of nutrients to plants. In particular, nitrogen, phosphorus and potassium are commonly applied, often in a mixed fertiliser which is referred to by the chemical symbols of its constituents, namely, NPK. The relative quantities of these three nutrients in the fertiliser are described by a simple formula. For example, a 20:10:10 fertiliser is one containing 20 parts of nitrogen, 10 parts of phosphorus and 10 parts of potassium. Differing ratios of N, P and K are obviously used to supply different amounts of each nutrient, according to the particular needs of the soil and crops.

The use of compound fertilisers of this type has increased considerably over recent years, and it is even common today to find small additions of trace elements (e.g. Cu, S or Mo) in the mixture in order to supply simultaneously the micronutrients needed by the soil. Nevertheless, many other types of fertiliser may be used to supply

**TABLE 4.5 SOIL pH BENEATH
TREES OF DIFFERENT SPECIES**
(from Ovington and Madgwick, 1957)

Tree	Average pH
Oak	7.64
Beech	7.61
Larch	7.46
Spruce	7.29
Pine	7.15
Alder	7.13

individual nutrients; for example potassium chloride (KCl) is used to provide potassium, and sodium nitrate ($NaNO_3$) to provide nitrogen.

It is important to note that the solubility of the different forms varies widely and, as a result, the elements supplied by the different types of fertiliser may not be equally available to plants. An important aspect of fertiliser management is raised here. **If the material is readily soluble the nutrients will rapidly become available to the plants**. This ensures a quick response. However, the plant may not need all the nutrients at one moment in its growth; instead it is likely to require a slow and continuous supply of nutrients throughout its growing period. Much of the fertiliser applied in this way may therefore be wasted; much, indeed, may be leached from the soil and enter stream waters where it is liable to cause pollution and eutrophication.

In order to ensure that fertilisers are used efficiently, it is clearly necessary to apply them in appropriate amounts and at an appropriate time. Fertilisers which are highly soluble need to be applied only a short time before the plants require them. Thus they are often spread at the same time as sowing, or afterwards, as the seedlings start to emerge. On the other hand, fertilisers which dissolve slowly are normally added some time before the crop is sown.

4.5 CASE STUDIES

4.5.1 Introduction

Because many of the chemical properties of the soil are relatively sensitive to changing environmental conditions, they provide ideal indices for analysing short-term variations in the soil. Measurements of both $CaCO_3$ content and pH for example have been used to illustrate the effects and rates of leaching in developing sequences of dunes (*Figure 4.4*), and on reclaimed land. Changes in pH associated with the development of forest vegetation have also been studied by measuring pH in forest plots of different age. Similarly, comparisons of pH under a variety of trees have shown the influence of different species upon soil acidity (*Table 4.5*).

Soil-topography relationships may also be investigated by chemical analyses. In the study of a catena sequence, for example, Furley (1968) showed that pH changed

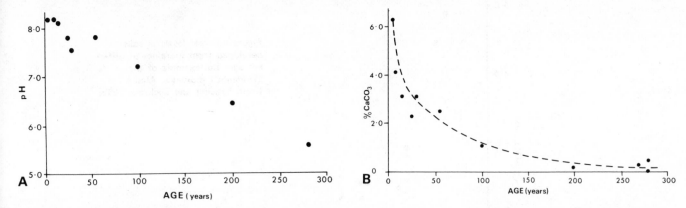

Figure 4.4 Changes in the pH and calcium carbonate contents of dune soils over time (from Salisbury, 1925)

4.5.2 Chemical properties and soil chronosequences (Crocker and Dickson, 1955)

significantly in relation to slope gradient. Because the soils on the steeper slopes are thin and intensely leached, pH is low. On low-angle slopes, however, accumulation of soil and nutrients results in increased pH levels (*Figure 4.5*).

The problem
Over time, the soil may be expected to become more acid. Metallic cations, such as calcium and magnesium will be leached and replaced by hydrogen ions derived from dissociation of the soil water or from plant roots.

The nitrogen content of the soil will similarly change over time. In this case, however, the general trend will be for an increase in the nitrogen levels of the soil. This is because nitrogen is not supplied by weathering, but by fixation and cycling by plants and soil organisms. Freshly weathered, immature soils contain very little nitrogen since organic activity is limited. With time, however, the vegetation and associated soil organisms will evolve and nitrogen will accumulate in the soil.

The aim of this study was to test these hypotheses of soil development in an area of glacially derived soils. In addition, the intention was to obtain data to quantify the rate of soil development under controlled climatic conditions.

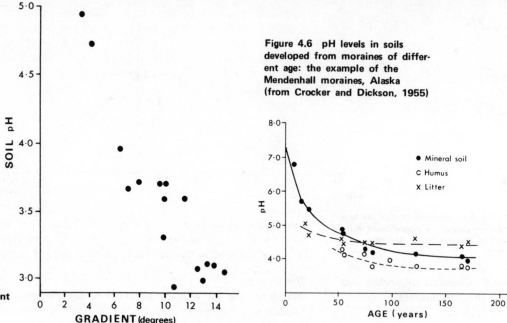

Figure 4.6 pH levels in soils developed from moraines of different age: the example of the Mendenhall moraines, Alaska (from Crocker and Dickson, 1955)

Figure 4.5 The relationship between soil pH and slope gradient in soils developed from chalk (from Furley, 1968)

The method

Two study areas were selected in south east Alaska, one comprising the recessional moraines of the Herbert Glacier, the other the similar moraines of the Mendenhall Glacier. The two areas were about 16 km apart. Both sets of moraines form ridges 4.5–6.0 m in height, and are composed predominantly of siliceous igneous rocks (diorite). The moraines vary in age from about 5 years, close to the present glacier snout, to about 200 years (dating of the moraines was based upon survey evidence and tree-ring analysis).

Samples were collected from pits to a depth of about 60 cm. The samples were then dried and taken to the laboratory where they were passed through a 2 mm sieve to obtain the fine earth. The pH and total nitrogen contents of the samples were then measured. The nitrogen results were expressed on the basis of the soil volume, after calculating its bulk density.

Results

Analysis of the pH of the soil from three depths — the surface litter, the underlying humus and the upper part of the mineral soil — showed a similar trend in acidity in both valleys. During the first 10–20 years the pH of the mineral soil fell rapidly to a value of about 5.5; subsequently the rate of change was considerably slower and by about 150–200 years the pH had more or less stabilised at 4.5–5.0. Initially, the pH of the litter and humus layers were somewhat lower than the mineral soil, but in both areas these horizons showed a much slower decline in pH and stability had been reached within 100 years. By this time the pH of the surface litter averaged about 4.5–5.0, while the humus had a pH of about 4.0 (*Figure 4.6*).

Comparison of the pH variations down the soil profile at 4 sites in the Herbert valley and five sites in the Mendenhall valley also showed similar trends (*Figure 4.7*). Initially, the soil was slightly acid throughout the profile, the pH at the surface being only slightly lower than that at depth. With time the upper part of the soil became more acid. The greatest fall in pH occurred in the upper 15 cm of the profiles. At the base of the profiles there appeared to be a slight increase in pH, presumably owing to deposition of calcareous materials leached from the upper horizons.

In contrast, the nitrogen contents of the soils tended to increase over time (*Figure 4.8*). Again, it seems that the change was most marked during the first 20 years of soil development, and by 100–150 years equilibrium had been reached at a nitrogen content of about 2.5 kg m^{-2}.

The distribution of nitrogen throughout the profile changed markedly over this period (*Figure 4.9*). Early in the sequence, the distribution was fairly uniform, with only a slight concentration in the topsoil. With time, however, the concentration

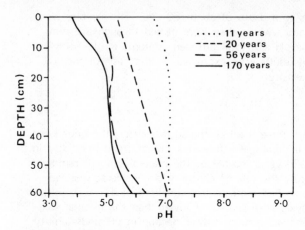

Figure 4.7 Variations in soil pH in profiles developed from moraines of different age: the example from the Mendenhall moraines, Alaska (from Crocker and Dickson, 1955)

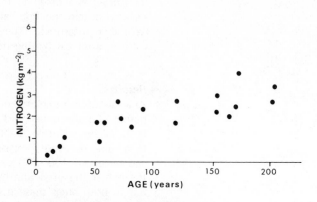

Figure 4.8 Nitrogen levels in soils developed from moraines of different age (from Crocker and Dickson, 1955)

in the upper 16 cm increased considerably, apparently as a result of the accumulation of organic matter at the soil surface.

Conclusions

It is clear from the results shown in *Figures 4.6* to *4.9* that the initial hypotheses were confirmed by this investigation. The pH does decrease over time, while the nitrogen content increases. It is interesting to note the rates of change displayed by the moraine soils, however. In all cases the major modification of the original material occurred during the first 15–25 years of soil development. The pH, for example, fell about 2 pH units during this period, yet only a further 0.5–1.0 units during the remaining 150 years or so. This pattern of change is, in fact, fairly typical of most soil properties. Development tends to be most rapid when the

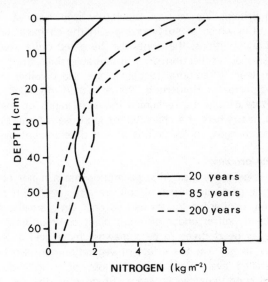

Figure 4.9 Variations in the soil nitrogen levels in profiles developed on moraines of different age: the example of the Herbert moraines, Alaska (from Crocker and Dickson, 1955)

disequilibrium between the soil and its environment is at a maximum. As the soil approaches equilibrium, the rate of change progressively falls.

The changes throughout the profile are also symptomatic of more general conditions. The topsoil, which is more intimately associated with the developing vegetation characteristically shows the most marked changes over time. One implication of this is that changes in soil properties over time are not solely a result of more prolonged weathering processes. To some extent they reflect a change in the intensity or character of weathering and soil development owing to the establishment of a different vegetation cover. We must remember that as the soil develops, so too does the vegetation; the two are intimately related.

Comment

The opportunity for carrying out studies of this type is somewhat limited, at least in Britain. As we have previously seen, however, dune soils provide an ideal situation in which to study changes in the chemical properties of the soil. Under suitable conditions, the dunes can be dated quite accurately, using evidence of old maps, personal recollection or, with suitable skills, from the age of the vegetation developed on them. Temporal sequences are also possible in many reclaimed areas, such as the warplands of Holderness, the reclaimed moorlands of the Pennines and Welsh uplands, (*Plate 6*) and the reclaimed industrial lands of coal-mining areas. Almost anywhere, in fact, where the character of land use, of soil type, or of the local environment has changed, investigations of soil development over time can be carried out.

4.5.3 Soil chemical properties beneath a tree canopy (Zinke, 1962)

The problem

It is generally clear that vegetation has a major effect on soil formation. On a large scale, particular types of soil are associated with prairie grassland, deciduous woodland, coniferous forests and so on. On a smaller scale it is apparent that soils frequently differ under different plant communities. We generally expect that the soil below beech trees to be somewhat more acid and less well mixed than those beneath oak trees. Yet the effect of vegetation can also be discerned at an even more detailed level. Beneath the canopy of an individual tree, quite marked variations in soil conditions can be seen. Frequently, the soil immediately adjacent to the trunk is bare and eroded, lacking in any accumulation of leaf litter. In addition, the chemical properties of the soil vary beneath the canopy. The aim of this investigation was to explain these patterns by studying detailed variations in soil chemistry around a single tree.

The method

A single pine tree (*Pinus contorta*), 45 years in age, was selected in an area of dune sand in western California. Samples were collected from the mineral soil at a depth of 7 cm, at distances of 0.3, 1.3, 2.5, 3.8 and 5.1 m from the trunk. Sample points were spaced out along four orthogonal transects, two of the transects being aligned parallel to the dominant wind direction. The soil samples were analysed for pH and nitrogen content.

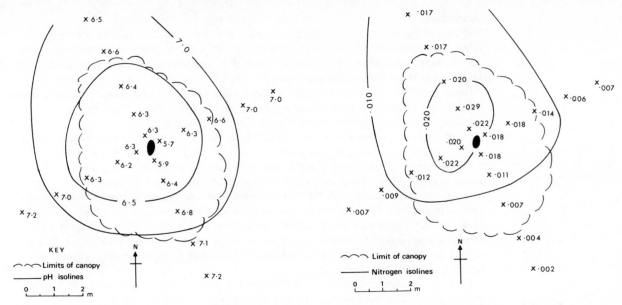

Figure 4.10 The pH of surface soil below a pine tree (from Zinke, 1962)

Figure 4.11 The nitrogen content (%) of surface soil below a pine tree (from Zinke, 1962)

Results

A plot of the pH results (*Figure 4.10*) illustrates clearly the influence of the tree and its canopy upon the chemical properties of the soil. Close to the tree trunk the soil was moderately acid, with pH values of less than 6.0. At a distance of about 0.5 m, however, the pH rose to about 6.3 and these values were characteristic of the rest of the area beneath the canopy. At the edge of the canopy the pH values rose again, to about 6.6–6.8, while beyond the canopy the average pH was 7.0–7.2.

The nitrogen content of the soil showed an almost inverse pattern (*Figure 4.11*). Nitrogen values were highest near the centre of the canopy and declined with distance away from the trunk. The rate of decline was greatest towards the south-east — the direction from which the main rain-bearing winds were derived.

Conclusions

The overall nature of the pH results suggests that the pine tree had an acidifying effect upon the soil. This is widely accepted, for it seems that coniferous trees produce organic substances which, when dissolved in rain-water, produce weak acids. These leach the soluble nutrients from the topsoil.

Conversely, the tree increased the nitrogen content of the underlying soil, apparently because the nitrogen is released during decay of the tree litter. Thus, the areas receiving most litter displayed the highest nitrogen contents.

If we look at the results in more detail, however, a number of other factors become clear. In the distribution of pH levels, a marked effect of the wind can be seen. The soil was generally more acid towards the north-west, downwind in relation to the main storm tracks. Thus, the acidifying influence of the tree seems to be extended downwind, presumably owing to litter and acidified rain-water being blown in this direction. The relatively low pH adjacent to the tree trunk is also interesting. Although this may partly reflect the greater incidence of leaching here, as a result of the concentration of water flowing down the tree trunk, it seems likely that the effect owes more to the natural acidity of the bark. In general, bark litter has a lower pH and a lower nutrient content than leaf litter.

Looking at the distribution of nitrogen in the soil, we can again see the influence of the wind. The zone of high N contents extends downwind (i.e. to the south-west) apparently because of the spread of litter in this direction. Additionally, there is some evidence that the nitrogen levels are at a maximum at about 1.0 m from the tree trunk. Once more, this seems to reflect the less nutritious character of the bark litter and, perhaps, the increased leaching of the soil by stemflow.

Comment

Several interesting points arise from this investigation. In the first instance, we may note the fact that the character of the litter derived from a single tree may vary considerably. This reflects the variations in nutrient accumulation and in organic composition of the different parts of the plant. In general, nutrient contents are highest in leaf material and lowest in the stem or trunk. At the same time, leaves tend to have a higher protein and sugar content, and lower lignin content, than the stem material. As we shall see in the next chapter, this influences the rate of decomposition and the release of nutrients into the soil.

The effect of the tree upon interception and throughfall of rainfall is also significant. In general, two areas of increased rainfall occur beneath a tree. In the area around the tree trunk rainfall is increased by the flow of water down the trunk (stemflow); in the zone peripheral to the canopy rainfall is increased by the shedding effect of the leaves — thus a canopy drip zone can also be identified. Both areas have an effect on soil properties, either by encouraging leaching or by concentrating nutrients which are washed from the leaves and atmosphere.

This case study also illustrates the importance of relatively subtle variations in soil properties. Frequently, these provide an extremely useful means of studying the processes of soil formation and the relationships of the soils to their surrounding environment. We have demonstrated this in relation to the effects of a single tree, but similar patterns can be seen under other plants. Variations in pH may be mapped under *Calluna* bushes, for example; subtle chemical variations may also be found alongside hedges. Clearly, numerous situations exist in which the detailed effects of vegetation upon soil conditions can be examined.

CHAPTER 5 BIOLOGICAL PROPERTIES

5.1 INTRODUCTION

5.1.1 The soil ecosystem

The biological component of the soil is extremely varied. It consists of both plant material and animals, living and dead. It includes the **macroflora**, the **macrofauna** and innumerable, diverse species of **micro-organisms** (*Figure 5.1*). Yet, despite the diversity of the soil biota, all these constituents are intimately interrelated.

The main process linking them is the transfer of **energy** through the soil. Energy is derived ultimately from the sun, but it is fixed within the soil by **autotrophic organisms** — mainly green plants which convert solar energy into a material form by the process of **photosynthesis**. This energy is then transferred through the soil, passing from one organism to another, along what are known as food webs. Some organisms, for example, obtain their energy directly from the autotrophic organisms, by eating the living plants. These **herbivores** may themselves be eaten by other organisms (carnivores), and thus the original energy fixed by the plants is moved through the soil system. Organisms which operate in this way — consuming the organic matter and digesting it internally — are known as **consumers**. Other organisms, however, operate more subtly. These release **enzymes** which decompose the organic matter outside their own bodies; then, when the organic matter is partly broken down, they absorb the decomposition products. These organisms are therefore called **decomposers**. They are the most diverse and numerous inhabitants of the soil, and include **fungi**, **bacteria** and **actinomycetes**.

A particularly important function of some decomposers is that they transform organic matter into forms in which it can be re-used by living plants. Nitrogen, for example, is released from the decaying plant and animal debris in the form of ammonia, but is converted by bacteria first into nitrite, and then into nitrate. In this form it can be dissolved in the soil water and taken up by plants. The process thus converts nutrients from unavailable to plant-available forms.

The close relationship between the various soil organisms therefore involves not only a transfer of energy, but a **cycling of nutrients**. These processes are the very basis of

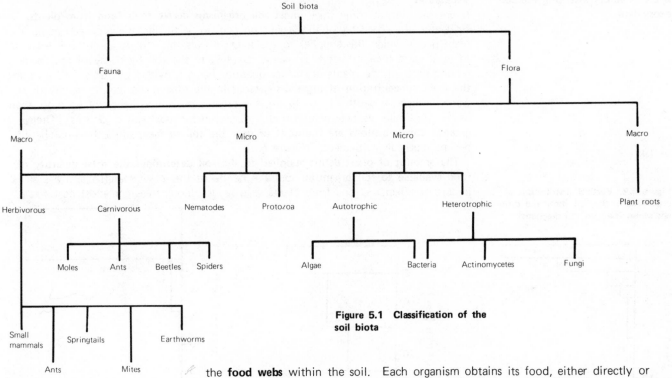

Figure 5.1 Classification of the soil biota

the **food webs** within the soil. Each organism obtains its food, either directly or indirectly, from neighbouring organisms. And, each time one organism eats or decomposes another, a transfer of energy and of plant nutrients takes place. This simplified picture of the soil organisms is, in fact, typical of almost all ecosystems. Only the names of the characters are changed! Thus, the whole complex association of organisms in the soil represents what is often known as the soil ecosystem.

5.1.2 Factors affecting the soil ecosystem

Vegetation

It is clear that, at some stage, **most soil organisms derive their food from plants.** Consequently, the character of the vegetation and, in particular, the type and amount of plant debris which this supplies to the soil, controls the activity of the soil organisms. In most cases organic matter is mainly supplied to the soil by decay of the above-ground parts of the plants, a smaller quantity being provided by roots. As a result, **the main concentration of organic matter is in the topsoil,** decreasing rapidly downwards until, at a depth of 20–30 cm, the organic content is generally no more than 3–4%. A similar vertical distribution is displayed by most soil organisms. **Their greatest concentrations are found at or near the soil surface,** while their numbers decline markedly with depth (*Figure 5.2*).

The amount of plant debris supplied to the soil determines the total quantity of food available to soil organisms, but the nature of this plant material also has an important effect. Debris from plants such as deciduous trees and most pasture

Figure 5.2 Vertical distribution of organisms in the soil (note the different scales used on the diagrams)

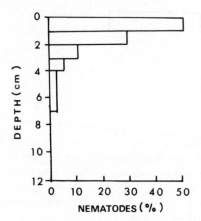

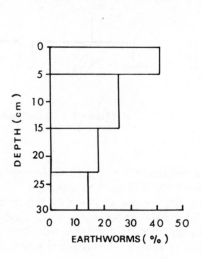

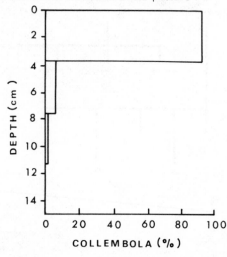

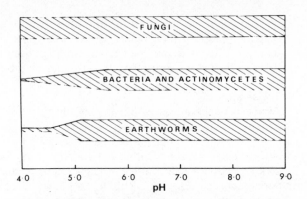

Figure 5.3 The effect of pH on the activity of soil organisms

grasses (e.g. *Fesctuca* and *Poa*) support a varied and active population of organisms. Coniferous trees and heathland vegetation, such as ling (*Calluna*), heather (*Erica*) and bracken (*Pteridium*), however, tend to be associated with a very restricted range of organisms. To some extent this difference seems to be due to the higher nutritional value of the former type of vegetation; it contains greater quantities of nutrients in a more easily obtainable form than the coniferous and heathland plants. In addition, these differences seem to relate to the effects of soil acidity, and this, therefore, is the second major factor influencing the biological composition of the soil.

Acidity
Most soil organisms are tolerant of a restricted pH range. Optimum populations occur at or near neutrality (pH 6.0–7.5); in acid to very acid conditions (below pH 4.5) the activity of soil organisms is generally very restricted (*Figure 5.3*). The main exceptions to this are certain species of fungi which are most prolific in acid conditions.

The exact nature of the influence of acidity is not clear; pH certainly has an independent effect, but it is also closely related to the nutritional status of the vegetation. Acid conditions give rise to nutrient deficient vegetation which is a poor source

145

of food for most organisms; on the other hand, neutral and alkaline soils generally support more nutritious plants, which decompose to provide an abundant food supply. Thus, it is the close relationship between pH and vegetation which is particularly effective in controlling the size and distribution of the soil population.

Temperature

Although temperatures within the soil do not vary as much as air temperatures, extremes of heat or cold nevertheless occur and are detrimental to most soil organisms. The greatest fluctuations occur in the top few centimetres of the soil, particularly when it is bare. During hot summer days the direct solar radiation on to a dark soil surface may cause the temperature in this surface layer to reach $35°C$; during cool, clear winter nights it may fall well below freezing point. Under a vegetation cover, or deeper in the soil, the fluctuations in temperature are considerably reduced. In unvegetated soils, however, the upper few centimetres generally support an impoverished soil fauna.

Light

Light is of most immediate importance to plant growth, and thus it mainly affects the soil organisms indirectly, by controlling the supply of plant debris to the soil. However, it also governs the rate of photosynthesis of algae. These are therefore concentrated in the upper part of the soil profile; few algae occur below a depth of 10–15 cm.

Aeration and drainage

Most soil organisms require a plentiful supply of oxygen and, at the same time, a humid environment. Consequently, **the balance between the moisture and air content of the soil is critical**. This is governed largely by the volume and size of the pore spaces. In soils with few macropores, waterlogging tends to occur and the soil organisms are inadequately supplied with oxygen. On the other hand, where soils are excessively well drained, the lack of moisture in the soil throughout much of the year leads to dehydration of many organisms. Similarly, within the top few

Plate 8. Peat, Roxburghshire, Scotland. Note the dark, undifferentiated profile and the well-developed angular structure as the peat dries out (the vertical cuts at the surface have been made during peat excavation) (Photo D. Briggs).

centimetres of most cultivated soils, the rate of evaporation is such that the soil tends to be too arid for organisms to survive.

5.2 TECHNIQUES

5.2.1 Organic matter content

Introduction

As we have seen the organic matter in the soil contains a wide variety of constituents. Despite its diversity, however, it is valid in many cases to treat it as a single component. Not only are the various constituents closely related, but they also operate upon the soil in a similar and consistent manner.

One of the major effects of the organic matter is upon the structural properties of the soil. Roots, earthworms and many micro-organisms exude organic compounds which help bind the soil particles, and thereby produce stable aggregates. Organic matter also influences the chemical fertility of the soil. Decomposition releases nutrients which, through the action of various soil organisms, are transformed into plant-available forms. The humus which develops as an intermediate product of this decomposition also acts as a store for nutrients.

Organic matter occurs in a number of forms in the soil. Under conditions of rapid accumulation and slow decomposition it builds up as a distinct surface layer. If this layer is thicker than about 50 cm it is normally called **peat** (*Plate 8*). Under less extreme conditions, some mixing and decomposition may take place with a result that the organic topsoil is less distinct. Where decomposition and mixing are only partial, the humus is left as partly decomposed masses within the mineral soil. It is then referred to as **discrete humus**. Where breakdown and mixing are complete, however, it is impossible to recognise the humus as a separate entity — at least without a microscope — and the humus is said to be **intimate**. Thus a sequence can be seen which reflects the balance between the rate of organic matter supply and the rate of decomposition and mixing (*Table 5.1*). This sequence can be used as a basis for the field description of organic matter. Nevertheless, in most cases a measure of the quantity of organic matter in the soil is more meaningful.

Measurement

Equipment: Porcelain crucible
     Oven
     Furnace
     Dessicator containing calcium chloride or silica gel
     Tongs
     Heat-proof glove

Sample: 5–10 g soil

Procedure: Dry the soil in the oven at 105 °C for 24 h; cool in a dessicator.

     Weigh the crucible and record the weight as W_1.

     Weigh out approximately 5 g soil into the crucible; weigh the crucible plus the soil and record the weight as W_2.

     Place the crucible in the furnace at 400–450 °C for 16 h.

TABLE 5.1 DESCRIPTIVE MEASURES OF ORGANIC MATTER CONTENT

Peat	Organic matter in a distinct horizon >50 cm deep
Mor humus	Acid organic matter in a distinct surface horizon, less than 50 cm deep
Mull humus	Neutral-alkaline organic matter, intimately mixed with mineral matter, in a distinct surface horizon, less than 50 cm deep
Discrete humus	Separate patches and lumps of humus within a mineral horizon
Intimate humus	Humus intricately intermixed with mineral matter so that individual particles of organic matter are not visible

Using tongs and the heat-proof glove, remove the crucible; cool in a dessicator; weigh the crucible plus its contents and record weight as W_3.

Calculation: The organic matter (OM) content is given by the loss in weight during ignition, as a percentage of the initial sample weight:

$$OM\ (\%)\ = \left(\frac{W_2 - W_3}{W_2 - W_1} \right). 100$$

The simplest method of measuring the organic content is to ignite the soil at high temperature. The organic matter is then burnt off (oxidised) and the loss of weight gives a measure of the organic content.

The main problem with the loss-on-ignition technique is that material other than the organic matter may be destroyed. Water held within the lattice structure of the clay particles, and carbonates such as $CaCO_3$ and $MgCO_3$ may be lost in this way. Consequently, over-estimation of the organic content may occur, particularly in clayey or calcareous soils.

In order to minimise these effects slow ignition (for 12–16 h) at 400–450 °C is recommended for most purposes. However, in some circumstances more rapid ignition (e.g. 2 h at 600 °C or 30 min at 800 °C) may be used. In these cases, however, over-estimation of the true organic matter content may occur; certainly these more rapid methods should not be used in very clayey or calcareous materials.

An alternative approach is possible where a furnace is not available. The sample is placed in a crucible, partially covered with a lid, and heated over a bunsen burner. Heating is continued until the sample ceases to fume, then the crucible is removed, cooled in a dessicator and the weight loss determined as above. Although this method is very rapid, it is less accurate than using a furnace since the temperature of ignition cannot be closely controlled.

5.2.2 Earthworm population

Earthworms serve several functions in the soil. They ingest and decompose plant litter. They mix and overturn the soil, and thus help to maintain weathering by repeatedly introducing soil materials to new weathering environments. They produce gums which stabilise the soil aggregates. They produce wormcasts which are not only structurally very stable, but also contain concentrations of many nutrients. They create channels which act as pathways for air and water movement. Small wonder, therefore, that they are often called 'the farmer's friends'.

They are also affected by many different factors. Most earthworms are tolerant only of slightly acid to alkaline conditions (*Figure 5.3*); thus they are rare in acid soils. They require a well-aerated, yet moist, environment, and consequently are absent from waterlogged soils. Temperature, too, is critical, and they are unable to survive excessive heat or frost. Finally, they depend upon an adequate supply of organic matter, for this is their main source of food. In general these requirements are most adequately met in the topsoil, and therefore most earthworms are found in the upper 20–30 cm of the soil (*Figure 5.4*). However, during periods of drought, frost or intense heating (e.g. when cereal stubble is burnt), they may migrate to much greater depths.

The numbers of earthworms in the soil vary considerably according to these conditions. Nevertheless, in many soils the numbers are remarkably high. One study in

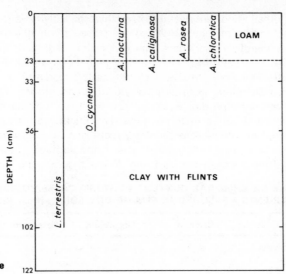

Figure 5.4 Vertical distribution of different species of earthworm in the soil (from Satchell, 1967)

Kent produced an estimate of 6 million per hectare, weighing almost 1700 kg, and the average for pasture land seems to be in the order of 1.25 million. In arable soils they are less abundant, partly because they are killed by ploughing, partly because continued cultivation destroys the organic content of the soil, and thus their food supply.

The amount of soil which they move is also phenomenal. This movement results from the transport of the soil to the surface where it is deposited in wormcasts. Charles Darwin estimated that this process caused overturning of as much as 50 tonnes of soil per hectare each year — enough to produce a layer 0.5 cm thick. Overturning by earthworms is therefore an extremely significant process; it is, for example, responsible for the production of the well-mixed mull humus which develops in many neutral and alkaline soils.

Another important feature of earthworm activity is the production of worm casts. These consist of material which has passed through the stomach of the worm, and which therefore contains a high proportion of organic gums. These gums give the casts a very high degree of stability and thus breakdown of the casts during rainfall is extremely slow. In addition, the casts apparently contain a concentration of calcium, magnesium, potassium, phosphorus and nitrate (*Table 5.2*). Consequently, they are an important aspect of soil fertility. Casting is by no means a universal characteristic of earthworms, however. Of the ten species commonly found in Britain, only two — *Allolobophora longa* and *A. nocturna* — normally produce casts at the surface; most other species cast below ground. Moreover, earthworm activity is essentially seasonal, worms being dormant during the summer.

TABLE 5.2 CHEMICAL CONTENT OF WORM CASTS COMPARED WITH ADJACENT SOIL (NUTRIENTS MEASURED IN PPM OF DRY SOIL) (from Russell, 1973)

	Calcium	Magnesium	Potassium	Phosphorus	pH
Wormcast	2790	492	358	67	7.0
Arable soil (0–15 cm)	1990	162	32	9	6.4
Wormcast	3940	418	230	9	5.3
Forest soil	747	140	138	7	4.6

Measurement

Equipment: Spade
Polythene sheet and polythene bags
Garden sieve

Procedure: Quickly and quietly dig a square hole, *c.* 30 cm wide and 30 cm deep.

Carefully sieve the soil on to the plastic sheet, extracting the earthworms as they appear and placing them in the polythene bags.

Calculation: Estimate the surface area of the hole (A) and the number of earthworms collected (N).

The number of earthworms per hectare furrow slice (i.e. in one hectare to plough depth)

$$= \frac{N \times 10^4}{A}$$

(if the hole is 0.3 $\times$ 0.3 m the area = 0.09 m^2, i.e. 9 $\times$ 10^{-2} m^2);

The major difficulty in estimating the numbers of earthworms in the soil is to extract a representative sample of worms. One method which is commonly used is to flood the soil with dilute potassium permanganate (1.5 g l^{-1}) or formalin solution (5 ml of 40% formalin l^{-1}). These are applied at a rate of 5–7 l m^{-2}. This approach is unsatisfactory for three reasons. It does not ensure full recovery of the earthworms, and two or more applications may be necessary to drive deep burrowing worms to the surface. The depth to which the method is effective is unknown and therefore the volume of soil from which the worms have been extracted cannot be measured. Thirdly, the method is one which pollutes the soil, and unless the worms are washed in water they quickly die.

Consequently a preferable technique is to hand-sort a sample of soil. This can be carried out in the field, using a garden sieve to break up the soil aggregates. With care almost 100% recovery of earthworms is possible.

One other approach to the assessment of earthworm populations is to count or weigh the earthworm casts which occur within a defined area. An area approximately 2 m $\times$ 2 m is marked out and carefully cleared of all existing casts. The site is then revisited after a few days and the earthworm casts counted or, preferably, removed, dried and weighed. If desired, the chemical properties of the casts can also be analysed. As has already been mentioned, however, not all species of earthworm produce surface casts, and thus this method does not give a measure of the total worm population. Nevertheless, it is an excellent means of monitoring seasonal or short-term changes in earthworm activity without the necessity of disturbing the soil to any significant extent.

5.2.3 Arthropods

Introduction

Arthropods include a wide variety of animals, such as spiders, centipedes, millipedes, mites and beetles. Their role in the soil is equally varied. Centipedes and millipedes, for example, help decompose plant residues by macerating the organic matter; ants and beetles aid aeration by mixing the soil, by burrowing and building their nests and colonies; some beetles, mites and spiders act as predators, helping to maintain a balance in the soil community; other beetles, particularly in the larval stage, are herbivorous and attack the roots of living plants. It is almost impossible to isolate the activity and importance of any single species of arthropod, for they are all closely related and interdependent. Nevertheless, variations in both the total population and the relative numbers of the different species may be significant in relation to general soil conditions, and thus extraction of arthropods to allow population counts is often useful. In this way the factors governing the activity of the arthropods can be identified, the close relationships between different species can be studied and the effects of insecticides on the arthropods can also be investigated.

Measurement

Equipment: Bench lamp with shade and 60 W bulb
Filter funnel
Gauze square
250 ml flask
Methanol
Retort stand and clamps
Filter paper
Hand lens or microscope

Sample: Core of loose soil

Procedure: Set up the equipment as in *Figure 5.5*, standing the fresh soil core on the gauze square above the filter funnel; pour *c.* 100 ml 70% methanol into the flask and insert it below the funnel.

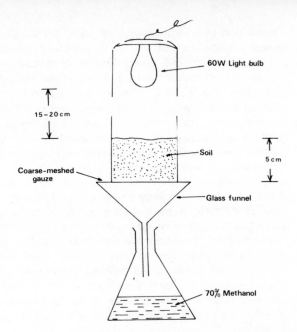

60W Light bulb

15–20 cm

Soil

5 cm

Coarse-meshed
gauze

Glass funnel

70% Methanol

**Figure 5.5 The Tullgren funnel
for extraction of arthropods**

Place the lamp *c.* 15–20 cm above the soil core and switch on the light.

Leave for 1–4 days, then filter the contents of the flask and collect the arthropods in the filter paper.

Using a hand lens or low-powered microscope identify the arthropods to genera level.

Dry soil and weigh; record the weight.

Calculation: Express the quantity of each genus as a frequency per 100 g of soil.

Extraction of arthropods is most easily carried out using a modified **Tullgren Funnel**. A thin core of soil (*c*. 10 cm deep) is heated from above by a light bulb, and this drives the arthropods downwards through the soil. As the soil dries out they are forced into a funnel and from there fall into a preserving solution of methanol. They can then be collected, studied under a microscope and identified.

Although this method is very simple, and is effective in loose soils through which the animals may easily move, it is of limited use in compacted soils. Recovery can be aided by breaking up the soil by hand in order to aid movement of the arthropods, but this inevitably introduces a source of inconsistency if comparative studies are being carried out. A second difficulty is the positioning of the lamp. If it is too close, the soil dries out too quickly and the arthropods die before they can migrate downwards. If the lamp is too far away the soil may take many days to dry out and drying may occur at the base of the soil core as well as at the top; the arthropods will then be trapped in the centre of the core.

Despite these drawbacks the technique is widely used and seems to give reasonably consistent recovery rates, at least for specific soil types. Once the arthropods have been extracted, however, the problem is to identify them. A limited key for this purpose is given in *Figure 5.6*, which illustrates some of the more common genera. For more detailed analysis, however, the text by Andrews (1973) or the charts provided by the British Ecological Society (see Appendix 2 for address) are recommended.

5.3 BIOLOGICAL PROCESSES

5.3.1 Organic matter decomposition

Organic matter, whether it is in the form of leaves and twigs falling on to the soil surface, or plants decaying *in situ*, consists of four main components: **carbohydrates, proteins, cellulose and lignin**. These compounds vary in their resistance to decomposition. Carbohydrates, for example, are least resistant and therefore decay most rapidly. Proteins are somewhat more resistant and thus break down more slowly, followed in turn by cellulose and lignin. Plants differ in the relative proportions of these components that they contain, and consequently organic material from different plants decomposes at different rates. For this reason debris from plants such as pine

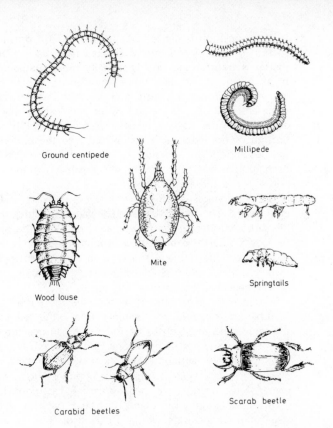

Figure 5.6 Identification chart for selected soil organisms

Ground centipede

Millipede

Mite

Springtails

Wood louse

Carabid beetles

Scarab beetle

trees (*Pinus*), fir trees (*Abies*) and ling (*Calluna*) decays very slowly, whereas most grasses, herbaceous plants and the leaves of broad-leaved trees (e.g. oak, elm etc.) decompose more rapidly.

Decomposition is carried out mainly by soil organisms, though chemical processes such as oxidation and hydrolysis also take place. **During decomposition, the organic matter is transformed into chemically more simple forms**; large complex molecules break down into smaller, more simple ones. Some of these decomposition products are released in gaseous form and are thus lost to the atmosphere (e.g. as CO_2 or H), some are dissolved in the soil water and lost by leaching (e.g. ions of Ca^{2+}, Mg^{2+} and K^+). Water itself is another common decomposition product. But a large proportion of the material is assimilated into the bodies of the organisms carrying out the decomposition. In time, of course, these organisms also die and are themselves decomposed, and thus, finally, the organic matter is completely broken down and released into the soil.

The early stages of this decomposition process in fact occur before the plant is even dead. The living tissues are invaded by **fungi** which decompose the most soluble carbohydrates (in particular, sugar). Once the plant is dead, new organisms take over. **Earthworms** and **arthropods** digest the plant debris and mix the macerated remains with the mineral soil. Fungi and **heterotrophic bacteria** attack the remaining carbohydrates; polysaccharides and starches are thus decomposed by what are known as the 'general purpose micro-organisms'.

Subsequently organisms of a more specific nature become active. Various species of fungi and bacteria, for example, attack the cellulose in the organic debris. At the same time the remains of the soil organisms themselves start to accumulate as the preceding waves of herbivores and decomposers, as well as their predators, die off. Ultimately, **actinomycetes** increase in numbers, for these are able to break down the most resistant organic compounds — the cellulose and lignin — as well as many of the products formed by earlier decomposition processes (*Figure 5.7*).

Organic material from plants such as broad-leaved trees and grasses reach this stage of decomposition within a period of 8–9 months. In the case of the more acidic, less nutritious vegetation, such as pine trees or heather, decomposition may take as long as 9 years. At the end of this period there remain only the most resistant lignins, together with the remnants of dead soil organisms and the more stable decomposition products (mainly **humic acid** and **fulvic acid**). These varied

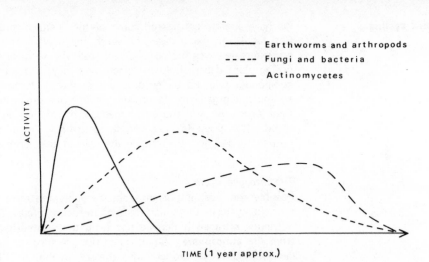

Figure 5.7 The activity of soil organisms during decomposition of organic matter

compounds occur as a dark brown or black amorphous substance known as **humus**; the process of decomposition to this stage is therefore called **humification**. It is the humus which gives topsoils their characteristic dark brown colour, and it is this material also which, by combining with clay particles and by adsorbing nutrients, plays such an important part in controlling soil structure and chemistry.

The humus, however, is not the end-product of decomposition. It merely represents a relatively stable phase, during which decay proceeds more slowly, at a rate of about 2% per year. Again, soil organisms play a vital role, but purely chemical processes of oxidation are of increasing significance. Because of the slow rate of decay, some of the humus in the soil may be extremely old. Using radio-carbon methods to date the material, it has been shown that the mean age of the humus can be as much as 1000 years in some soils, while ages of 400 years are not uncommon. Thus, humus found in present day soils may represent the decomposition products of plants sown in medieval times!

5.3.2 Nutrient cycling

We have already considered some of the chemical processes of nutrient cycling in Chapter 4. In many cases, however, the movement of nutrients through the soil ecosystem involves biological processes. As we have seen, soil organisms, in decomposing plant debris, help to release and recycle plant nutrients. In addition, the compounds released by decomposition may be further ingested, stored and liberated by the soil organisms themselves. Numerous organisms are involved in each stage of these processes, and the exact mechanisms of nutrient cycling are imperfectly understood. It is possible, however, to identify the main features of the most important nutrient cycles and to indicate the role of the various groups of soil organisms.

The nitrogen cycle

Possibly the most important — and certainly the most frequently studied — biological nutrient cycle in the soil is that of nitrogen (*Figure 5.8*). Unlike most plant nutrients, **nitrogen is derived not from the weathering processes within the soil, but from the atmosphere**. Small quantities are supplied by lightning, and somewhat larger amounts are brought into the soil in rainfall. The majority, however, is accumulated in the soil by micro-organisms.

Two main groups of organisms take part in this **fixation** process. One group — the **symbiotic nitrogen fixers** — live in close and mutually beneficial association with plants. This group consists mainly of bacteria, in particular various species of *Rhizobium*. These apparently infect the roots of certain plants (mainly legumes such as clover, but also a limited range of other plants), and lead to the formation of cyst-like **nodules**. The bacteria live within these nodules and are able to accumulate and fix nitrogen, which is then absorbed by the plant.

The second group of nitrogen fixing organisms live independently within the soil. They are thus known as **non-symbiotic** or **free-living** nitrogen fixers. They include a relatively wide range of bacteria, fungi and algae, of which the bacteria are generally the most important. Under aerobic conditions non-symbiotic fixation is mainly carried out by bacteria known as *Azotobacter*; in less well aerated anaerobic soils *Clostridium* is more active. Since most soils contain zones of poor aeration

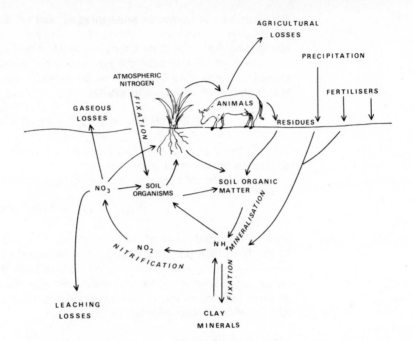

Figure 5.8 The nitrogen cycle

(e.g. within the peds) together with aerobic areas (e.g. in the larger pores and fissures), it is likely that both organisms are normally active.

In the case of symbiotic fixation, the nitrogen is accumulated within the host plant. Thus, release of the nitrogen into the soil occurs mainly when the plant dies and decays. Non-symbiotic fixation, on the other hand, takes place within the bodies of the soil organisms, the nitrogen being released upon the death and decomposition of these organisms. In both instances the nitrogen is released mainly in the form of **ammonium salts** (NH_4^+). This process of release during organic matter decomposition is known as **mineralisation**, and numerous organisms take part, including bacteria, actinomycetes and fungi. In effect, these organisms excrete excess nitrogen derived from the plant debris on which they feed.

In the form of ammonia, the nitrogen is not readily available to plants, and further alteration is necessary before it can be re-used. The first stage in this alteration is the oxidation of ammonia to **nitrite**;

$$2NH_4^+ + 3O_2 \rightarrow 2NO_2^- + 2H_2O + 4H^+ + energy$$

This is carried out by a limited range of autotrophic bacteria, of which *Nitrosomonas* appears to be the most important.

The nitrite formed by this reaction is rapidly oxidised further to produce **nitrate**:

$$2NO_2^- + O_2 \rightarrow 2NO_3^- + energy$$

Again a restricted range of bacteria take part in this process, in particular *Nitrobacter*. In the form of nitrate, the nitrogen can be absorbed by plants. However, nitrate is extremely soluble and can readily be leached from the soil by percolating waters. Hence, at this stage, considerable losses of nitrogen may take place. The side effects of these leaching losses, if prolonged, are the accumulation of nitrogen in stream and lake waters and their eventual eutrophication.

Other nutrient cycles

Although nitrogen is the nutrient most frequently referred to in reference to the activities of soil organisms in nutrient cycling, many other examples can be quoted. In the phosphorus cycle organisms are involved in both the conversion of unavailable, inorganic phosphorus to available forms, and in the return of organic phosphorus to the soil. Organisms also play a major part in the cycling of carbon. During organic matter decomposition carbon from plant residues is released as carbon dioxide; at the same time living organisms and plant roots respire CO_2 into the soil atmosphere. As a result the carbon dioxide content of the soil air is constantly being replenished by the processes of decomposition and respiration. A balance is maintained through the diffusion of CO_2 into the atmosphere.

In all these examples soil organisms provide an essential link between the uptake of nutrients by plants and their release back into the soil. Where organisms are scarce, the rate of nutrient cycling is reduced, and the availability of nutrients to the plant is limited. Thus, soil organisms ultimately control the rate of supply of nutrients to plants, and it is clear that the fertility of the soil depends upon the maintenance of a varied and active fauna within the soil.

5.4 AGRICULTURAL IMPLICATIONS

5.4.1 Biological aspects of soil fertility

Both organic matter and soil organisms are of considerable importance in relation to soil fertility. Through their effects on soil structure, they control the stability, aeration and drainage of the soil; through their role in nutrient cycling they influence the supply of nutrients to plants. For these reasons the maintenance of the biological processes is a critical part of soil management. Nevertheless, it is apparent that many modern farming methods alter these processes, occasionally with detrimental effects upon soil fertility.

Physical effects

The main physical implications of both organic matter and micro-organisms are related to the development of soil structure. As we saw in Chapter 2, the aggregation of soil particles is largely a result of electrochemical forces between colloidal particles. These attract the particles towards each other. Humus, like clay particles, has a well-developed surface charge and consequently contributes to this process of aggregation.

The stability of the aggregates formed by these forces is also dependent upon biological processes. In the first place plant roots and the minute thread-like **hyphae** of fungi enmesh the aggregates. Secondly, **gums** and **slimes** exuded by living roots and organisms, and released by organic decomposition, act as cements. Since these are insoluble in water they provide an extremely persistent stability. As a result, soils rich in organic matter tend to be structurally more stable than purely mineral soils.

Chemical effects

It is clear that the biological processes of organic matter decomposition and nutrient cycling are extremely important in the retention, release and re-use of plant nutrients. Two nutrients, in particular, are dependent upon biological cycling: nitrogen and carbon. All other nutrients, however, are at some stage taken up by plants and released during decomposition, thus being dependent upon biological processes of plant decay. This applies both to the macro-nutrients such as phosphorus, potassium and calcium, and to the micro-nutrients, such as iron, molybdenum, boron and copper.

Organic matter also acts as a vital store of available nutrients. As a result of decomposition, the humus attains a negative surface charge, similar to that possessed by clays. In the case of organic matter, however, the charge arises from the presence of relatively unstable compounds of carbon, oxygen and hydrogen. This charge allows nutrients to be adsorbed on to the humus colloids, where they are protected to some extent from leaching. Nevertheless, the nutrients held in this way are readily available to plants and may be released either by cation exchange or during further decay of the humus.

Several other chemical effects of the biological components of the soil should be noted. The first is the role of soil organisms in nutrient uptake by plants. As we have seen, fungi and bacteria may be active in assimilating nutrients and supplying them to the plants; nitrogen and phosphorus are particularly affected in this way. Conversely, soil organisms may reduce the supply of certain elements. The availability of nitrogen, for example, may be temporarily reduced during the early stages of organic decomposition, owing to uptake by the rapidly expanding population of bacteria. Until these start to die and decay, the nitrogen is withdrawn from the 'pool' of available nutrients in the soil. Competition for oxygen may also occur under anaerobic conditions, while, in situations of limited nutrient availability, certain symbiotic organisms may turn parasitic and attack their host plants. Finally, of course, some organisms are naturally parasitic, and by damaging plant roots reduce the ability of the plants to absorb nutrients.

5.4.2 The consequences of cultivation

Effects of tillage

Cultivation affects the biological properties of the soil in a number of ways. Possibly the most direct effect is that **ploughing kills many organisms**, particularly earthworms. More indirect influences occur owing to the increased aeration which results from tillage. In the short term this increases the activity of soil organisms by increasing the oxygen supply. These organisms decompose the plant debris and, in time, lead to depletion of the organic content; at this stage the soil population is reduced by the limited food supply. This process is simultaneously encouraged by the increased rate of chemical oxidation resulting from the improved aeration.

Under arable cultivation **the reduction in the amount of organic matter in the soil is also accelerated by crop removal**. Whereas the return of plant debris to the soil is reasonably constant in a natural system, where man harvests the crop there is a reduced input of organic matter. This tendency is most marked where root crops are grown, for in these cases almost the whole of the crop is removed, and with it all the nutrients bound up in the plants. Under grazing systems, the effect is less important, for much of the material removed is returned in the form of animal residues.

The overall effect of farming is therefore to reduce the quantity of organic matter in the soil. Today, in many areas of continuous arable cultivation, the organic content of the soil is below 3–4%; whereas under grassland or woodland vegetation organic matter contents form 10% or more of the soil weight. This loss or organic matter has a number of implications. It reduces the stability of the soil structure, it reduces the ability of the soil to hold moisture and nutrients, and it diminishes the rate of nutrient cycling.

Management of organic matter

In order to maintain the supply of organic matter, a variety of methods are used. In some cases it is possible to supply organic matter directly, in the form of **farmyard manure** (FYM). This in fact serves a dual purpose, for it decomposes to form a humus-like substance which gives the soil stability, and at the same time contains nutrients needed by crops. From the agricultural point of view, however, there are several disadvantages in using manure. In the first place it is bulky and costly to transport. Secondly, it is difficult to judge the quantities necessary to supply individual nutrients since the chemical composition of the manure is highly variable and often unknown. Thirdly, under modern conditions many farms do not produce sufficient manure to supply their needs.

An alternative approach is to **conserve** the organic matter in the soil as far as possible. This can be done by the use of **grass leys**, that is, by occasionally returning arable land to pasture. The lack of tillage, the lack of crop removal and the high productivity of the grass all help to replenish organic matter lost during previous years of arable cultivation (*see* Section 5.5.3). Over a three- or four-year period a substantial increase in the organic matter content and, through this, the population of soil organisms, may occur. This in turn leads to increased structural stability (*see Figure 2.10*).

In recent years several other methods of conserving organic matter have been devised. One approach is to leave as much plant residue on the surface as possible after harvesting. This technique of 'stubble mulching', as it is known, fails to conserve the organic matter completely, for some of the crop is still being removed. In addition, stubble mulching may encourage disease because the stubble provides shelter in which pests can live until the next crop emerges.

Another recent development is the use of zero-tillage methods, in which the soil is never — or, at least, only rarely — ploughed. Instead, the crop is sown directly into the untilled soil using a machine which cuts a narrow slit. When the crop is harvested the stubble is left in position and thus helps to retain the organic matter. As we have seen, however, zero tillage involves several problems; in particular, weeds and pests need to be controlled by use of herbicides and pesticides, while crop yields in some cases seem to be reduced (*see* Chapter 2).

5.5 CASE STUDIES

5.5.1 Introduction

The methods described in this chapter can be used to study a wide range of situations. Numerous studies have been carried out in an attempt to investigate the effect of various factors upon the population of soil organisms. Earthworms, in particular, offer

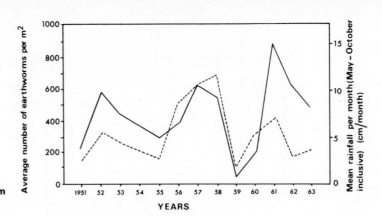

Figure 5.9 Relationships of earthworm numbers (solid line) to average monthly rainfall (broken line) (from Satchell, 1967)

scope for this type of approach. Repeated sampling of earthworms in a Welsh pasture, for example, showed a close relationship between the numbers of earthworms and the mean monthly rainfall (*Figure 5.9*). This indicates the importance of moisture upon earthworm activity; in general earthworms attempt to avoid dessication. Shorter term, seasonal changes, can also be studied in this way, and the relationship between earthworm numbers and air or soil temperature examined. Similarly, spatial variations in the population of individual organisms have been analysed, and attempts made to relate numbers to environmental factors. Thus pH has been shown to be an important control upon certain species of earthworm (*Figure 5.3*). Comparisons of single soil types under different types of cultivation also show the influence of land use upon earthworm populations (*Figure 5.10*).

Various studies have also been based upon the measurement of organic matter content. Seasonal variations in the organic matter content of the soil have been related to the supply of organic matter during autumn (when the organic content is at a maximum) and the rate of decomposition (mainly in the spring and summer). Similarly, variations in the organic content of soils under different types of land use have

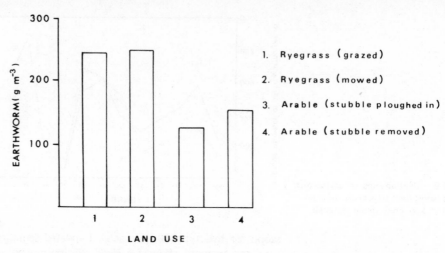

Figure 5.10 Earthworm numbers under different cropping systems (from Heath, 1962)

1. Ryegrass (grazed)
2. Ryegrass (mowed)
3. Arable (stubble ploughed in)
4. Arable (stubble removed)

been analysed. However, studies of soil processes, such as changes in specific properties over time, often require the standard techniques of analysis to be developed or modified. Thus our first detailed case study considers some simple adaptations which can be applied to the measurement of organic contents in order to study processes of organic matter decomposition.

5.5.2 Analysis of organic matter decomposition
(Edwards and Heath, 1963)

The problem
It is generally accepted that the main processes of organic matter decomposition are carried out by micro-organisms, particularly bacteria and fungi. The role of the larger organisms, such as earthworms, is more obscure, though it has been suggested that they are important in preparing the vegetable matter for decay by macerating the debris and by carrying it down into the soil, where the micro-organisms can attack it.

The aim of this study was to investigate the role of the various organisms by monitoring the decomposition of plant material under controlled conditions.

The method

Leaves of oak and beech were collected and discs, 2.5 cm in diameter, cut from them. Fifty discs of each type of leaf were then placed in a series of nylon mesh bags, 10 cm X 7 cm. Each bag had a different mesh size, thus allowing entry to a restricted range of soil organisms.

Mesh size	Organisms with free enetry
7.000 mm	All micro-organisms, and invertebrates
1.000 mm	All organisms except earthworms
0.500 mm	Micro-organisms, mites, springtails and small invertebrates
0.003 mm	Micro-organisms only

The bags were buried at a depth of 2.5 cm in newly cultivated soil. After one month they were recovered and the contents studied to determine the area of leaf that had disappeared through decomposition. To achieve this, the leaf discs were compared with a perspex disc 2.5 cm in diameter, divided into 100 squares. This was placed over the leaf disc and the area of decomposed leaf assessed by counting the squares. The bags were then returned to the soil and the procedure repeated every 2 months for a further 8 months.

Results

The general pattern of results was similar for both types of leaf, in all the bags, though the 7.0 mm and 0.5 mm meshes showed the most consistent trends (*Figure 5.11*). Decomposition was most rapid during the first five months, after which a decline in the rate of decay occurred. The discs of oak leaves, however, were consistently decomposed more rapidly than the beech leaves. Even more significantly, decomposition was greatest in the bags with the largest mesh; within a period of 9 months about 95% of the oak leaves had decayed in the 7.00 mm bags, and about 72% of the beech leaves. By contrast, only 40% and 35% respectively of the material in the 0.5 mm bags had been decomposed during this time. There is also evidence to suggest that the most rapid phase of decomposition occurred later in the smaller bags. In the 7.00 mm mesh bags, for instance, maximum decay

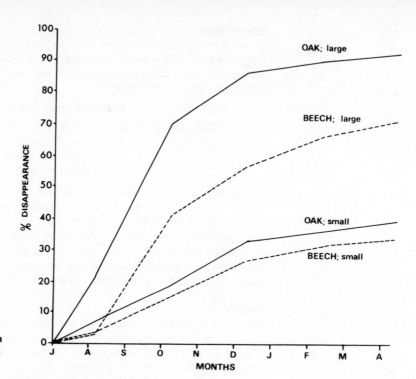

Figure 5.11 The decomposition of leaf discs by soil organisms: large 2.0 mm mesh; small 0.5 mm mesh (from Edwards and Heath, 1963)

took place between one and three months. On the other hand, the rate of decomposition in the 0.5 mm bags reached a peak between three and five months.

Conclusions

Several conclusions can be drawn from these results. Firstly, it is apparent that the larger arthropods and, in particular, earthworms, play a significant role in organic matter decomposition, though the exact nature of their effect is not shown by this experiment. Secondly, the effect of these macro-organisms occurs early during decomposition, for it is during the first 3–5 months that the greatest differences between the various meshes occurred; after this time the rate of decay was similar for all the bags.

It is similarly apparent that the main difference between the oak and beech leaves took place during the early phases of decomposition. The rate of decay of the beech leaves was markedly less than that of the oak leaves during the first month, but after that little difference in the speed of decay was discernible. In addition, the greatest disparity between the decay curves of the two leaf types occurred in the coarsest mesh bags; in the smaller bags the rates of decay were almost identical. For these reasons it seems likely that the relative resistance of the beech leaves to decay arises from their reduced attack by earthworms. This may relate to a dietary preference of the worms for oak leaves.

Comment

The approach used in this study can be adapted to a variety of problems. Edwards and Heath suggested for example that beech leaves are more resistant to decay than are oak leaves, and it is generally agreed that coniferous material takes longer to decompose than does deciduous plant debris. Clearly it would be possible to test these assumptions by comparing the rates of decomposition of different types of leaves using the methods devised by Edwards and Heath. Similarly, it would be possible to compare the rates of breakdown of single species of plant debris in different site conditions; for example, in acid and alkaline soils, or in cultivated and uncultivated land.

This approach can, however, be simplified in a number of ways. One modification is to weigh the organic matter in the bags, rather than to measure changes in surface area. This requires that the moisture content of the leaves is determined on each occasion by drying and weighing a small sample of material. If desired a separate, duplicate sample can be used for this purpose so that drying does not interfere with the decomposition processes.

5.5.3 Grass leys and soil organic matter (Clement and Williams, 1964)

The problem

The use of grass leys in order to 'give the soil a rest' from continuous cultivation has long been practised by farmers, but the ways in which the grass crop benefits the soil is not clear. It appears (as we saw in Section 2.5) that the stability of the structure improves under grass, and it has also been shown that the nitrogen content of the soil increases. Both changes, however, seem to be indirect responses to the establishment of a grass crop; the critical factors are apparently changes in the biological properties of the soil.

This study was designed to examine the effects of ley grass upon the quantity of organic matter in the soil. Although the original project considered numerous aspects of grassland management, we will discuss here only the general differences between grassland and arable cultivation.

The method

Two sets of field plots were set up. One set was sown with rye-grass and clover, the other was used for the production of wheat and barley. Soil samples were taken from each plot six months after the start of the experiment, and then annually for a period of six years. The samples were collected in cores, 2.5 cm in diameter and 15 cm deep; thirty samples from each set of plots were taken on each occasion, and these bulked to give an average sample for the whole set. The organic matter content of each sample was measured, and expressed in terms of percentage organic carbon (this is approximately 75% of the total organic matter).

After a period of three years a subsidiary set of samples was taken from the grass plots. These consisted of core samples at two centimetre intervals down the profile to a depth of 12 cm. These were analysed to show the vertical distribution of organic matter in the soil. Six months later further samples were taken at a depth of 10–12 cm in both the grass and arable plots in order to compare the effect of the two agricultural systems on the soil at depth.

Results

Over a period of six years a consistent trend was shown by the two sets of plots

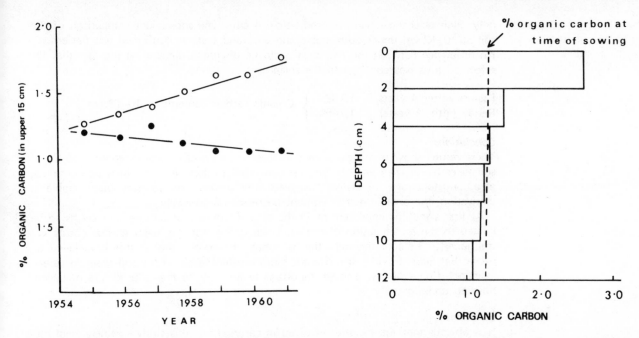

(*Figure 5.12*). Under grass, the soil displayed a slow increase in organic matter which, rising from about 1.25% in 1955 to 1.75% in 1961, involved a 40% increase over a period of 6 years. By contrast, the soil under continuous arable cultivation experienced a decline in the organic matter content. In 1955 the organic matter in the upper 15 cm averaged approximately 1.2%, by 1961 it had fallen to 1.0%, a decline of about 17%.

The measurement of the distribution of organic matter in the pasture soil three and a half years after sowing showed a marked concentration in the surface layers (*Figure 5.13*). While the organic content in the upper 2 cm had almost doubled,

Figure 5.12 Changes in the organic matter content of soil under different cropping systems (from Clement and Williams, 1964)

Figure 5.13 Vertical distribution of organic matter in the soil after a 3½ year grass ley (from Clement and Williams, 1964)

only minimal changes had occurred below 4 cm. The subsequent comparison of the soil at 10–12 cm under both arable and grassland systems confirmed this pattern. No difference between the two soils could be discerned, illustrating the fact that the effect of grass was confined to the surface:

Pasture (after 4 years)   1.132% ⎫
Barley (after 4 years)   1.145% ⎬ Organic carbon contents at 10–12 cm
                                 ⎭

Conclusions

These results show that a significant increase in the organic matter content of the soil takes place when an arable soil is converted to pasture. This increase continues, at a consistent rate, for at least six years after sowing. By contrast, under arable cultivation, the organic matter content decreases progressively.

At first sight this improvement in the organic content of pasture soils seems to be related to the accumulation of root and leaf debris, for the major change occurs in the upper 2–3 cm of the soil — the root mat. However, other studies have failed to prove that grass provides significantly more organic debris to the soil than do cereal residues. Consequently, a more important factor may be the reduced rate of decomposition under grass.

Comment

It is obvious that the facilities involved in carrying out this study — namely field plots cultivated for chosen crops, and a period of six years in which to make the measurements — are not normally available. Yet this does not mean that similar experiments are not possible. In many cases it is feasible to replace the time scale with a spatial one, by analysing fields in different phases of a rotation. Thus, two adjacent fields may be found, on the same soil type, but under different crops. Each can be intensively sampled and differences in the organic matter content determined. It is also possible to use the same approach to study other aspects of the soil ecosystem. Thus numbers of earthworms or arthropods under different agricultural systems can be monitored.

The investigation considered above concentrates upon the effects of different agricultural systems. The same approach can, however, be used to study the effects of afforestation. At the simplest level a direct comparison can be carried out between soils under forest and adjacent agricultural soils. At a more sophisticated level, organic matter contents or numbers of soil organisms can be measured in forested plots planted at different times. Most commercial plantations contain a range of plots of different age and they provide suitable conditions in which to study progressive changes in the biological properties of the soil resulting from changes in land use. Indeed, with a little imagination, many other situations can be thought of which would provide similar opportunities for studying the effects of land use on soil biology.

APPENDIX 1: CONSTRUCTION OF THE DIP-WELL PROBE

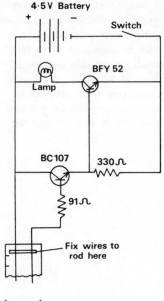

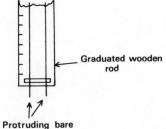

Graduated wooden rod

Protruding bare ends of wires

Equipment

Wooden rod, *c.* 1 m in length.
333Ω resistor.
91Ω resistor.
4.5 V battery.
2.5 W bulb.
Transistor (BC107).
Transistor (BFY52).
Single copper flex, insulated.
Flex clips.
Araldite.

Procedure

Mark off the wooden rod in 0.5 cm intervals.

Wire up the sensor unit as shown in the diagram opposite; this is best housed in a protective wooden or plastic box.

Run the two ends of the flex down the wooden rod, fixing in position where necessary with Araldite or flex clips.

Cut the flex so that the two ends protrude *c.* 3 mm below the bottom of the rod; bare the ends of the flex and fix in position, 3–5 mm apart, with Araldite.

Check that the bulb lights up when the probe is inserted in water.

Use

The instrument should always be dried between readings; if the light stays on after the probe has been removed from the water, check that no water is bridging the bared ends of the flex.

Always switch off the instrument when not in use since a slow leakage is possible through the transistors.

For further details, see Darlington, E. (1971) *Journal of Agricultural Engineering Research*, **16**, 423–4

APPENDIX 2: USEFUL ADDRESSES

The following firms carry a wide range of equipment for soil studies; catalogues can normally be obtained on request.

Gallenkamp,
P.O. Box 290,
Technico House,
Christopher Street,
London EC2P 2ER

Griffin & George Ltd.,
Ealing Road,
Alperton,
Wembley,
Middlesex, HAO 1HJ

Walden Precision Apparatus,
Shire Hill,
Saffron Walden,
Essex, CB11 3BD

As well as from Gallenkamp and Griffin & George, chemicals can be obtained from:

British Drug House,
Chemicals Ltd.,
Poole,
Dorset, BH12 4NN

For various maps of soils and related information, the following organisations are particularly useful:

> Soil Survey of England and Wales,
> Rothamsted Experimental Station,
> Harpenden,
> Herts, AL5 2JQ

> Soil Survey of Scotland,
> The Macaulay Institute for Soil Research,
> Craigiebuckler,
> Aberdeen, AB9 2QJ

> Land Utilisation Survey,
> Miss Alice Coleman,
> King's College,
> Strand,
> London, WC2R 2LS

For practical experience in the use of field and laboratory techniques, contact:

> Field Studies Council,
> Information Officer,
> Preston Montford,
> Montford Bridge,
> Shrewsbury, SY4 1HW

This organisation runs a variety of courses at centres throughout the country for both teachers and students.

> British Ecological Society,
> Harvest House,
> 62 London Road,
> Reading, RG1 5AS

GLOSSARY

ADHESION	Molecular attraction between two surfaces, e.g. water and a solid particle.
ADSORPTION	The attraction of ions or compounds to the surface of a solid, as a result of electrochemical forces.
ANAEROBIC	The condition in which oxygen is deficient; for example in waterlogged soils.
ANION	A negatively charged ion (e.g. Cl^-, OH^-).
AUTOTROPHIC	Capable of using CO_2 or carbonates as the sole source of carbon and of obtaining energy from the oxidation of inorganic substances.
AVAILABLE WATER	The moisture within the soil which is available to plants, normally considered to be that held between a tension of 0.05 and 15 bars.
BULK DENSITY	The mass of dry soil per unit volume.
CAPILLARITY	The ability of liquids to rise in narrow tubes, against the force of gravity.
CATENA	A sequence of soils developed upon a slope resulting from the operation of slope-forming processes (e.g. creep and rainwash).
CATION	A positively charged ion (e.g. Ca^{2+}, Na^+).
CHRONOSEQUENCE	A sequence of soils developed over different lengths of time.
CLAY	Particles finer than 0.002 mm (2 μm) in diameter.
COHESION	The attraction of like molecules to each other, e.g. between two clay particles.
COLLOID	Very small particles which have a net surface charge.
CONSISTENCE	The resistance of the soil to crushing or compaction.
DISSOCIATION	The tendency for molecules to split into separate ions.

DIVALENT	Referring to an ion with a valency of two (e.g. Ca^{2+}, O^{2-}).
ELUVIATION	The removal of soil material in suspension or solution by percolating waters; forms an eluvial horizon.
FIELD CAPACITY	The state of the soil when all the gravitational water has been removed by drainage, normally about 2 days after rainfall.
FRAGIPAN	A compact soil horizon, often a result of periglacial activity or of compaction by machinery.
FREEZE–THAW	The action of alternate freezing and thawing of water in the soil or rocks; expansion of the water on freezing causes disintegration.
GLEYING	The formation of dull grey and green colours in the soil caused by **reduction** under waterlogged conditions.
GRAVITATIONAL WATER	The water which drains from the soil under the influence of gravity, normally considered to be that held at tensions of less than 0.05 bar.
HORIZON	A layer in the soil formed by pedological processes; often designated by a letter, e.g. the *A* horizon (the surface layer of mixed organic and mineral matter); the *B* horizon (the subsurface layer of maximum illuviation); the *C* horizon (the weathering parent material).
HUMUS	Partially decomposed organic residues.
HYDRAULIC CONDUCTIVITY	The rate at which water flows through the soil.
HYGROSCOPIC WATER	Water adsorbed on to the surface of solid particles in the soil.
ILLUVIATION	The deposition of soil particles or solutes washed from overlying soil horizons by percolating waters; forms an illuvial horizon.
INFILTRATION	The movement of water into the soil from the atmosphere.
ION EXCHANGE	The interchange of ions between a colloidal surface and the surrounding solution, or between two colloidal surfaces.
KAOLINITE	An alumino-silicate clay with a 1:1 lattice structure.

LABILE	Stored, inert.
LEACHING	Removal of soil materials in solution by percolating waters.
LEY	A temporary pasture (pronounced *lay*).
LOAM	A mixture of **clay**, **silt** and **sand** in the approximate ratios of 1 part clay to 2 parts silt and 2 parts sand.
LOESS	Fine **sand** and **silt** deposited by the action of wind; apparently the material is derived from sediments ground up by glacial activity.
MACRO-NUTRIENTS	Nutrients required by plants in large quantities, including hydrogen, oxygen, carbon, nitrogen.
MATRIC FORCES	The forces of **cohesion** and **adhesion** which attract water molecules to soil particles and thus retain water in the soil.
MICRO-NUTRIENTS	Nutrients which are required by plants in small quantities, including copper, boron, molybdenum and zinc.
MONTMORILLONITE	An alumino-silicate clay with a 2:1 lattice structure; montmorillonite is able to absorb water and thus expands when wetted.
MOTTLES	Patches of different coloured material in the soil produced by alternate **oxidation** and **reduction**.
OXIDATION	The loss of an electron from an atom, which thus increases its positive charge; a reaction normally occurring in **aerobic** conditions and leading to the formation of red and yellow **mottles**.
PEAT	Soil formed of organic matter which builds up, normally under **unaerobic** or acidic conditions.
PED	An aggregate; a naturally occurring structural unit within the soil.
pH	A measure of the acidity of a solution; defined as the negative logarithm to the base of ten of the hydrogen ion concentration.

REDUCTION	The gain of an electron by an atom, which thus reduces its positive charge; the opposite of **oxidation**.
SAND	Particles between 2.0 mm and 0.06 mm in diameter.
SILT	Particles between 0.06 mm (60 μm) and 0.002 mm (2 μm) in diameter.
SLAKING	The detachment of soil particles from aggregates caused by the explosion of air bubbles trapped within the soil when it is wetted.
STRUCTURE	The arrangement of soil particles into aggregates.
TEXTURE	The 'feel' of the soil when it is worked in the hand; commonly used to define the distribution of particle sizes in the soil, in particular the relative proportions of **sand**, **silt** and **clay**.
TOPOSEQUENCE	See **catena**.
TURGIDITY	The swollen, resilient state of plants when well-supplied with water.
WETTING FRONT	The lower limit of the zone of moist soil which slowly penetrates into the soil after rainfall.
WILTING POINT	The state of the soil at the point where plants start to suffer from dehydration; generally defined as the moisture content at a tension of 15 bar.

REFERENCES

Andrews, W.A. (Ed.) (1973) *Soil ecology*, Prentice-Hall, Ontario

Briggs, D.J. (1977) *Sources and methods in geography: sediments,* Butterworths, London

Cannell, R.Q. and Finney, J.R. (1973). 'Effects of direct drilling and reduced cultivation on soil conditions for root growth', *Outlook on Agriculture,* **44**, 184–189

Chow, V.T. (Ed.) (1964) *Handbook of applied hydrology*, McGraw-Hill, New York

Clarke, G.R. (1951) 'The evaluation of soils and the definition of quality classes from studies of the physical properties of the soil profile in the field', *Journal of Soil Science* **2**, 50–60

Clement, C.R. and Williams, T.E. (1964) 'Leys and soil organic matter: 1. The accumulation of organic carbon in soil under different leys', *Journal of Agricultural Science, Cambridge* **63**, 377–83

Crocker, R.L. and Dickson, D.A. (1955) Soil development on the recessional moraines of the Herbert and Mendenhall glaciers, south-east Alaska, *Journal of Ecology* **43**, 169–85

Daugherty, R. (1974) *Science in geography: data collection*, Oxford University Press, London

Davies, D.B. and Cannell, R.Q. (1975) 'Review of experiments on reduced cultivation and direct drilling in the United Kingdom, 1957–1974', *Outlook on Agriculture* **46**, 216–20

Edwards, C.A. and Heath, G.W. (1963) 'The role of soil animals in the breakdown of leaf material'. In Doeksen, J. and van der Drift, J. (Eds), *Soil organisms*, pp.76–84, North Holland Publishing Company, Amsterdam

Furley, P. (1968). 'Soil formation and slope development. 2. The relationship between soil formation and gradient angle in the Oxford area', *Zeitschrift für Geomorphologie,* **NF12**, 25–42

Hammond, R. and McCullagh, P.S. (1974) *Quantitative techniques in geography*, Clarendon Press, Oxford

Heath, G.W. (1962) 'The influence of ley management on earthworm populations', *Journal of the British Grassland Society* **17**, 237–44

HMSO (1970) *Modern farming and the soil*, Ministry of Agriculture, Fisheries and Food

Low, A.J. (1955) 'Improvements in the structural state of soils under leys', *Journal of Soil Science* **6,** 179–97

McCullagh, P. (1974) *Science in geography: data use and interpretation*, Oxford University Press, London

Ovington, J.D. and Madgwick, H.A.I. (1957) 'Afforestation and soil reaction', *Journal of Soil Science* **8,** 141–9

Russell, E.W. (1973) *Soil conditions and plant growth*, Longman, London

Salisbury, E.J. (1925) 'Note on the edaphic succession in some dune-soils with special reference to the time factor', *Journal of Ecology* **13,** 322–8

Satchell, J.E. (1967) 'Lumbricadae'. In Burges, A. and Raw, R. (Eds), *Soil biology*, pp.259–322, Academic Press, London

Scott, V.H. (1956) 'Relative infiltration rates of burned and unburned upland soils', *Transactions of the American Geophysical Union* **37,** 67–9

Small, T.W. (1972) 'Morphological properties of driftless area soils relative to slope aspect and position', *The Professional Geographer* **24,** 321–6

Smith, R. and Atkinson, K. (1975) *Techniques in pedology*, Elek Science, London

Soane, B.D., Butson, M.J. and Pidgeon, J.D. (1975) 'Soil/machine interactions in zero-tillage for cereals and raspberries in Scotland', *Outlook on Agriculture* **46**, 221–6

Thomasson, A.J. and Robson, J.D. (1967) 'The moisture regimes of soils developed on Keuper Marl', *Journal of Soil Science* **18**, 329–40

Thow, R.F. (1963) 'The effect of tilth on the emergence of spring oats', *Journal of Agricultural Science, Cambridge* **60**, 291–5

Zinke, P.J. (1962) 'The pattern of influence of individual forest trees on soil properties', *Ecology* **43**, 130–3

FURTHER READING

GENERAL

Brady, N. (1974) *The nature and properties of soils*, MacMillan, New York

Courtney, F.M. and Trudgill, S.T. (1976) *Soils: an introduction to soil study in Britain*, Arnold, London

Kohnke, H. (1968) *Soil physics*, McGraw–Hill, New York

Russell, E.W. (1971) *The world of soil*, Fontana, London

TECHNIQUES

Andrews, W.A. (Ed.) (1973) *Soil ecology*, Prentice-Hall, Ontario

Clarke, G.R. (1971) *The study of soil in the field*, Clarendon Press, Oxford

Smith, R. and Atkinson, K. (1975) *Techniques in pedology*, Elek Science, London

Taylor, J.A. (1960) 'Methods of soil study', *Geography* **45**, 52–67

STATISTICAL METHODS

Daugherty, R. (1974) *Science in geography: data collection*, Oxford University Press, London

Davis, P. (1974) *Science in geography: data description and presentation*, Oxford University Press, London

Hammond, R. and McCullagh, P.S. (1974) *Quantitative techniques in geography*, Clarendon Press, Oxford

McCullagh, P.S. (1974) *Science in geography: data use and interpretation*, Oxford University Press, London

Siegel, S. (1956) *Nonparametric statistics*, McGraw-Hill, New York

PHYSICAL PROPERTIES OF THE SOIL

Low, A.J. (1954) 'The study of soil structure in the field and the laboratory', *Journal of Soil Science* **5**, 57–78

Low, A.J. (1973) 'Soil structure and crop yield', *Journal of Soil Science* **24**, 249–59

Russell, E.W. (1971) 'Soil structure: its maintenance and improvement', *Journal of Soil Science* **22**, 137–51

HYDROLOGICAL PROPERTIES OF THE SOIL

Curtis, L.F. and Trudgill, S.T. (1974) 'The measurement of soil moisture', *British Geomorphological Research Group Technical Bulletin* 13

Hills, R.C. (1970) 'The determination of the infiltration capacity of field soils using the cylinder infiltrometer', *British Geomorphological Research Group Technical Bulletin* 3

Reynolds, S.G. (1970) 'The gravimetric method of soil moisture determination', *Journal of Hydrology* **11**, 258–99

Thomasson, A.J. (1976) 'Soils and field drainage', *Soil Survey Technical Monograph* 7

CHEMICAL PROPERTIES OF THE SOIL

Ball, D.F. and Williams, W.M. (1968) 'Variability of soil chemical properties in two uncultivated brown earths', *Journal of Soil Science* **19**, 379–91

Pearsall, W.H. (1952) 'The pH of natural soils and its ecological significance', *Journal of Soil Science* **3**, 41–51

BIOLOGICAL PROPERTIES OF THE SOIL

Andrews, W.A. (Ed.) (1973) *Soil ecology*, Prentice-Hall, Ontario

Hayes, A.J. (1965) 'Studies in the decomposition of coniferous leaf litter. 1. Physical and chemical changes', *Journal of Soil Science* **16**, 121–40

Jackson, R.M. and Raw, F. (1966) *Life in the soil*, Studies in Biology, 2. Arnold, London

Satchell, J.E. (1958) 'Earthworm biology and soil fertility', *Soils and Fertilisers* **21**, 209–19

INDEX

Withdrawn